Earth in Flames

Earth in Flames

How an Asteroid Killed the Dinosaurs and How We Can Avoid a Similar Fate from Nuclear Winter

OWEN BRIAN TOON AND ALAN ROBOCK

OXFORD
UNIVERSITY PRESS

OXFORD
UNIVERSITY PRESS

Oxford University Press is a department of the University of Oxford.
It furthers the University's objective of excellence in research, scholarship,
and education by publishing worldwide. Oxford is a registered trademark of
Oxford University Press in the United Kingdom and in certain other countries.

Published in the United States of America by Oxford University Press
198 Madison Avenue, New York, NY 10016, United States of America.

CIP data is on file at the Library of Congress

ISBN 9780197799703

DOI: 10.1093/9780197799734.001.0001

Printed by Integrated Books International, United States of America

Brian dedicates his work on this book to his children—Katie, Will, and Russ—in hopes that they will live to see a time when the world no longer must worry about nuclear winter.

Alan dedicates his work to his children (Brian and Dan), his grandchildren (Danny, Genevieve, and Enzo), and to all the young people on Earth—it's up to you to rid the world of nuclear weapons so that we do not suffer the same fate as the dinosaurs.

Contents

PART III. EPILOGUE. COULD IT HAPPEN?

Preface: How We Met and a Brief History

Our story begins in 1980. Luis Alvarez, Nobel Prize in Physics winner and passenger on the Great Artiste chase plane when the nuclear bomb was dropped on Hiroshima, his son Walter, a geologist, and coworkers created an uproar in geology and paleontology by publishing data showing that a worldwide half-inch-thick layer of rock was debris from a large asteroid impact that occurred at just the same time as the mass extinction that killed the dinosaurs (Alvarez et al. 1980). Geologists at the time rejected any catastrophic theory for events in Earth history and did not like the Alvarez hypothesis. Paleontologists were aghast that a physicist might have solved their most cherished problem. As a result of the controversy, a meeting was held in Snowbird, Utah, in 1981 involving people from a wide range of scientific disciplines to assess the Alvarez hypothesis. Brian Toon presented a talk at the meeting showing that the dust raised in such an impact could blot out the sun and for six months to a year create freezing temperatures across the globe, making the Alvarez hypothesis a plausible mechanism to cause a mass extinction without killing everything on the planet (Toon et al. 1982). One of the meeting facilitators, Lee Hunt, who worked for the National Academy of Sciences, suggested to Brian that he should consider the parallel with nuclear wars.

When Brian returned to his work at the National Aeronautics and Space Administration (NASA) Ames Research Center, he contacted his colleague Rich Turco, who worked for R&D Associates, a think tank that studied nuclear weapons, among other things. Brian and Rich quickly found that indeed the dust raised in a nuclear war could cause a massive climate change. In parallel, Paul Crutzen, who later won the Nobel Prize in Chemistry, and John Birks suggested that burning forests in a nuclear war would produce a smoke pall so dense it would produce "Twilight at Noon" for several weeks (Crutzen and Birks 1982). Brian and Rich quickly realized that more smoke would come from burning cities than from forests and that the smoke would rise well above the surface, prolonging its effects. Working with colleagues Tom Ackerman and Jim Pollack, who also worked on the dinosaur extinction problem, they found a nuclear war could have devastating impacts on the environment rivaling those that caused the extinction of the dinosaurs (Turco et al. 1983).

As graduate students, Brian and Jim had worked for Carl Sagan, and they knew Carl was interested in nuclear wars because of the Drake equation, as he explained in his 1980 television series *Cosmos*. The Drake equation, used to compute how many intelligent civilizations might be present in our Milky Way

Galaxy, predicts that intelligent civilizations should be common. Carl hypothesized that the last term in the Drake equation, the lifetime of an intelligent civilization, must be short to explain why we had not observed other intelligent civilizations. Carl feared that young intelligent civilizations would develop nuclear weapons and annihilate themselves. So, because of his interest in the problem of nuclear conflicts, Brian and Jim invited Carl to join them. It was later pointed out to the group that the initials of the authors of their first paper— Turco, Toon, Ackerman, Pollack, and Sagan—sounded out TTAPS, the military funeral bugle call.

As Alan Robock perused the schedule of talks at the annual American Geophysical Union (AGU) Fall Meeting in San Francisco in December 1982, he noticed a very interesting title: "Climatic Effects of Nuclear War." But when he went to hear the talk, a notice on the door said the talk had been canceled. Three of the authors—Brian Toon, Tom Ackerman, and Jim Pollack—worked at the NASA Ames Research Center at the time. The center director worried that if NASA scientists studied the consequences of nuclear war, it could negatively affect the center's funding since nuclear weapons were not part of NASA's research areas as defined by Congress. So, he required a review of the work to make sure it was scientifically sound and apolitical before it could be published. Carl Sagan took this requirement as an opportunity to conduct a review involving much of the world community that worked on climate problems. To prepare for the review, a draft paper, referred to as the Blue Book, was written. NASA Ames also conducted an internal review that eventually required that any published work not use words such as "war" or "conflict." During the next few months as the team worked to complete a journal article on this topic, they had to come up with a new way to describe what they had found and called it "nuclear winter," following wording that Rich had used in the Blue Book to describe the effects of a nuclear war. This was such a powerful way of describing their discovery in just two words, the cause and effect (or in scientific jargon, the forcing and response), that it persists today as an iconic warning to the world.

Meanwhile, as the team worked on the final paper, they informed the scientific community of the work by circulating copies of the Blue Book as part of the review that Sagan organized. The review was held in Cambridge, Massachusetts, in April 1983 and involved almost 100 participants. Two Russians, Vladimir Aleksandrov and Georgiy Stenchikov, did their own calculations based on input from the Blue Book, using a three-dimensional climate model that was created in the United States, and got the same result: a nuclear war between the United States and Russia would cause a nuclear winter (Aleksandrov and Stenchikov 1983). Mike MacCracken, of the Lawrence Livermore National Laboratory, published a report on the climate effects of a nuclear exchange with a three-dimensional model (MacCracken 1983).

Alan went to the first open conference on nuclear winter, "The World after Nuclear War," held at the Shoreham Hotel in Washington, DC, in November 1983, where the matching US and Russian results were presented. This inspired him to employ an energy-balance climate model he had been using to study the impacts of volcanic eruptions to look at the long-term effects of nuclear war (Robock 1984), further reinforcing this result. All these climate models were primitive by today's standards, given the computer and scientific understanding limitations at the time, and so there were critics of the results. Stenchikov immigrated to the United States to work with Alan at the University of Maryland, and later at Rutgers University, and he contributed to our later work on nuclear winter.

Over the next few years, in various workshops and conferences, these shocking results were discussed, debated, and validated. A National Academy of Sciences report (National Research Council 1985) confirmed that nuclear winter was a valid theory, as did a study by the Scientific Committee on Problems in the Environment, a group representing national academies of science around the world. In 1990, the TTAPS team in a *Science* article (Turco et al. 1990), and colleagues Carl Sagan and Rich Turco in a book (Sagan and Turco 1990), summarized all this work as the Cold War ended and the threat of nuclear war seemed to recede.

These studies influenced the political process. Presidents Ronald Reagan and Mikhail Gorbachev, who each cited the nuclear winter studies, said in a joint statement in 1985 that a nuclear war "cannot be won and must never be fought." After years of negotiation, they agreed upon the 1987 Intermediate-Range Nuclear Forces Treaty, or INF Treaty. This landmark agreement led to the removal of about 2,200 intermediate and short-range ground-based missiles and launchers from Europe, effectively ending the nuclear arms race. The Soviet Union dissolved in 1991. It seemed that peace had broken out and that Russia and the West no longer saw each other as enemies. Alan and Brian went on with other research pursuits. We felt wonderful that we had played a part in making the world more peaceful, helping to end the nuclear arms race and bringing about a decline in the global nuclear arsenal. Figure P.1 shows Carl Sagan with Rich Turco and Brian Toon in 1994 when they presented talks honoring Carl on his sixtieth birthday (Terzian and Bilson 1997). Several important Russian academics were also present at Carl's birthday symposium, such as Georgi Arbatov, who was an advisor to five general secretaries of the Communist Party of the Soviet Union. He spoke at the symposium about a nuclear-free world.

During this interlude, Brian, Rich Turco, and their colleagues turned their attention back to the problem of asteroids impacting Earth. They investigated a wide range of sizes of asteroid and comet impacts and concluded that to have global extinction effects asteroids needed to be near the size of the one that killed

Figure P.1 Richard Turco, Carl Sagan, and Brian Toon at the Sagan Symposium celebrating Carl's sixtieth birthday in 1994 in Ithaca, New York. Rich presented a talk on nuclear winter at the symposium, while Brian spoke about the environments of Earth and other worlds. Photo by David Morrison

the dinosaurs. Data showing that soot was present in the geologic debris left behind by the dinosaur-killing asteroid led them to see that the effects of the asteroid impact were very similar to those predicted for nuclear winter.

Then at the 2005 AGU Fall Meeting, Brian and Rich Turco tracked down Alan in a hallway. Brian told Alan that he had been asked about the consequences of a nuclear war between two new nuclear states, India and Pakistan. After first thinking it would not produce significant global impacts, Brian and Rich now estimated that it could result in millions of tons of smoke in the stratosphere. Alan asked who was going to calculate the climate impacts. They said they thought he might do it. Thus began a 20-year-long (and continuing) collaboration to study the climatic impacts of nuclear war with modern climate models.

Even though we had little funding, we started to work on this problem and discovered that a nuclear war between India and Pakistan could produce climate change unprecedented in recorded human history. These climate changes were not cold enough to be a nuclear winter, but they would have devastating global impacts on agriculture. We also found, by using modern climate models, that the nuclear winter theory is indeed correct and that a nuclear winter could follow a war involving the United States, Russia, and their allies, likely killing the majority of the people on the planet. Our results helped lead to the International

Figure P.2 Nuclear winter pioneers Rich Turco, Brian Toon, Tom Ackerman, Alan Robock, and Gera Stenchikov at the American Geophysical Union Fall Meeting, December 2019

Campaign to Abolish Nuclear Weapons (ICAN) getting the Treaty on the Prohibition of Nuclear Weapons passed by the United Nations in 2017. As a result, the 2017 Nobel Peace Prize was awarded to ICAN "for its work to draw attention to the catastrophic humanitarian consequences of any use of nuclear weapons and for its ground-breaking efforts to achieve a treaty-based prohibition of such weapons." Again, we were proud of this, but we also realized that the problem of devastating environmental change due to nuclear conflict was still with us. And we needed to do more work to figure out, for various scenarios, how much smoke would get into the atmosphere, how much the climate would change, and how this would affect crops, fisheries, and, ultimately, the global food supply.

Professors can do research on any topic they want. It's called "academic freedom," and it's a wonderful thing. But these days you need more than a pencil and paper, or a laptop, to do your science. The work involves modifying and running complex computer programs that simulate how the climate system behaves, which requires the fastest supercomputers in the world. It involves supporting graduate students and postdocs by paying salaries, tuition, benefits, overhead to the university, publication costs, and travel to conferences (like the AGU Fall Meeting, where our latest results are presented, we get feedback on them, and we learn about what our colleagues are doing). Figure P.2 shows us at the last meeting we could attend before the COVID-19 pandemic started, at a poster session we organized on our most recent work (Witze 2020).

All this requires money, which we typically get by submitting grant proposals to US government agencies, such as NASA or the National Science Foundation. But we could not find program managers who would even consider proposals to do more nuclear winter work, not at these agencies nor at the Department of Energy (which makes the nuclear weapons), the Department of Defense (which manages our nuclear arsenal and makes plans to possibly use it), the Department of Homeland Security (responsible for protecting us from the possible effects of nuclear war), or the national intelligence community (whose job is to protect us from potential threats).

Surprisingly, there were no funds to support work on the reasons that the dinosaurs died either. While not a political hot potato like nuclear winter, studies of the dinosaur extinction event were limited to geological data collection. Funding agencies were tired of supporting studies of the causes of the extinction event. In addition, separate camps developed around the causes of the extinction that bitterly fought each other over the mechanisms. Peer review did not work well under these circumstances.

We continued to do a little work, tacking some simulations onto research supported for other purposes, such as to figure out the impacts of volcanic eruptions on climate, but we were limited in the time we could spend on nuclear winter. In one such study, Mike Mills, who had been one of Brian's postdoctoral students, discovered in 2014 that massive ozone loss could follow a nuclear war between India and Pakistan.

Then one day in 2017, Claire Zabel, a program manager for the Open Philanthropy Project, called Alan for some advice on a project they were considering funding related to stratospheric geoengineering, a topic he was working on. After they talked, he asked her if they would consider funding our work on nuclear winter. Claire was not familiar with this topic, so she asked for a five-page proposal, which Brian and he provided, with what we thought was an audacious $1.2 million budget for a 3-year project. After a quick review, she asked for a longer proposal that she would send out for external review and said to be sure to ask for more money if we needed it to ensure we got results in three years. We assembled a team to study the topic comprehensively, including establishing scenarios of how nuclear war might be fought between different nuclear states, assessing the fuel loads in cities and industrial areas that would be ignited by nuclear weapons, and modeling how those areas would burn and inject smoke into the upper troposphere. The group also simulated how the smoke would loft into the stratosphere, how the smoke would be transported around the world, the properties of the smoke, the climate response, the impact on ozone, the agricultural response to the changed climate, and how the world economic system would produce changed prices and affect the availability of food. We asked for $2.98 million and within a couple of months had the grant. We had never in our careers had such an experience. We got more money than we originally asked for—and in a record time. In our first three years of research, we produced a number of papers in high-impact journals, including *Science* and the *Proceedings of the National Academy of Sciences*, whose results only strengthened our fear of the possibility of a nuclear holocaust. In 2020, we were renewed for another three years for another $3 million.

Surprisingly, it was also difficult to find funding to work on the extinction of the dinosaurs, but there have been perhaps 100 times more studies of how the dinosaurs did die than how we might die. Fortunately, one of Brian's graduate students, Charles Bardeen, proved to be a computer genius. He created a version of the National Center for Atmospheric Research Community Climate Model that provided a leap forward in our ability to perform simulations of both nuclear wars and the dinosaur extinction. New simulations followed in 2017.

In 2022, Brian and Alan, among others, were awarded a prize by the Future of Life Institute for "reducing the risk of nuclear war by developing and popularizing the science of nuclear winter." More importantly, in 2023, the Future of Life Institute provided about $4 million for a small number of scientists in the United States and in other countries to investigate the dangers of nuclear war. Perhaps, at last, a wider community of researchers will develop, and they will warn their nations about the danger we all face from nuclear weapons.

As we continue our work, we feel an obligation to warn humanity about what we have discovered. Our scientific articles, popular press articles, tweets, and TED talks have only reached a limited number of people. For example, Brian's TEDx talk, which links the dinosaur extinction and nuclear winter, had been watched by 2024 by 8.8 million people, but that is only 0.1% of the world population. We have written this book to try to broaden the audience for the parallels we have found between the extinction of the dinosaurs and nuclear wars. We hope to motivate citizens to make their voices heard together to end the threat of nuclear weapons as soon as we can. As Beatrice Fihn, the executive director of ICAN said when she accepted the Nobel Peace Prize in 2017, "The story of nuclear weapons will have an ending, and it is up to us what that ending will be. Will it be the end of nuclear weapons, or will it be the end of us? One of these things will happen. The only rational course of action is to cease living under the conditions where our mutual destruction is only one impulsive tantrum away."

In this book, we explain how, unlike the dinosaurs that were killed by an unexpected asteroid impact, we know that a nuclear war might be coming, and therefore we have the power to eliminate this threat. We have found direct parallels between how the dinosaurs died and what might happen to us following a nuclear war. This is the story of how the dinosaurs died, how we might follow in their path, and how, as Carl Sagan feared, we could become an intelligent civilization that ended not with a whimper but with a bang.

Chapter 1
Prologue

It was a warm, humid, springtime day in paleo-Mexico during the Late Creta-ceous geologic period 66 million years ago. The sun shone bright in a cloudless sky, and the air was still and quiet. The dinosaurs—dominant among all life at the time—were emerging from their shady retreats to feed on the lush vegeta-tion. Suddenly, a blinding flash from over the nearby sea flooded the landscape, and the shadows vanished. The startled animals hardly had time to turn and look upward at the light. Then the most formidable beasts ever to roam the planet were quickly vaporized. Destructive earthquakes, tsunamis, and shock waves roared across the land and the oceans for thousands of miles, annihilat-ing everything in their path. Fires were ignited everywhere across the planet when debris from the crater that formed from the asteroid impact fell back to Earth. The day of dinosaur reckoning had arrived. Life on Earth would never be the same again.

It was a pleasant summer morning in Hiroshima, Japan, about 80 years ago. The sun was bright, and the air was still and quiet. People, the descendants of the mammals that inherited Earth from the dinosaurs, were attending school, tending their fields, working on their houses, and manufacturing goods and mil-itary equipment. Three aircraft flew high above, of little concern relative to the fleets of hundreds of aircraft that had been dropping incendiary bombs on other Japanese cities. Something fell from one of the planes and floated slowly down on a parachute. Suddenly a blinding, searing flash of light flooded the landscape within a mile and a half or so of ground zero near the prewar International Exhibition Center building. People were badly burned from the light flash. Then powerful shock waves knocked them and their houses down. City residents a few miles away, rushing to their windows to see what had happened, were stunned as glass shattered in their faces. The pilot of the aircraft that dropped the bomb, Colonel Paul Tibbets, saw fires springing up across the city. Within hours, the fires merged to form an immense firestorm. Winds flowed toward ground zero at 30 to 40 miles per hour, dragging debris inward toward a towering thunder-storm hovering over the city, raining wet, radioactive debris blackened by smoke down on the citizens. Life on Earth would never be the same again.

The dinosaurs, and about 75% of other species of animals that we know about, didn't die 66 million years ago across the Earth because they were hit

Earth in Flames. Owen Brian Toon and Alan Robock, Oxford University Press. © Oxford University Press (2025). DOI: 10.1093/9780197799734.003.0001

on their heads by pieces of an asteroid. The asteroid, about the size of Mt. Everest, plunged into the ground at a speed about 10 times faster than the fastest bullet from a rifle near what is now the small village of Chicxulub in the Yucatan Peninsula of Mexico. The impact left a crater in the ground about 180 km (110 miles) in diameter, now buried by sediments and largely covered by ocean. How could such an isolated event in one tiny corner of Earth, now marked by such an inconspicuous crater that it went unnoticed for many years despite active searches, lead to extinctions across the planet? The science community worked for more than 40 years to piece together what happened. While many theories were initially advanced, the surviving ideas all have in common a sudden, decades-long global environmental catastrophe caused by atmospheric aerosols, clouds of particles, that were injected into the atmosphere directly by the impact or indirectly by fires caused by the impact. These particles can still be found in geologic layers across the Earth, corresponding to the time of the mass extinction event, and it was through the discovery of these layers and fossils therein that the impact was first detected.

The extinction of the dinosaurs and the effects of nuclear conflicts are usually studied as separate topics. However, there are obvious parallels between nuclear wars and the asteroid impact because explosions of atomic weapons also inject particles into the atmosphere that can cause a global environmental catastrophe. Over time, we have discovered more and more common elements between asteroid impacts and nuclear war. We find that investigating these common elements reveals a lot about each of them. This book explores these parallels.

Part I of this book tells not just the story of how the dinosaurs died but also how other asteroid and comet collisions threaten us and what we can do to stop another impact. Part II describes how a nuclear war might produce fatalities across Earth, even outside of the combat zones, from similar mechanisms to those that killed the dinosaurs. The military is ignoring this danger. In our epilogue, Part III, we discuss the inevitable extinction of our species, *Homo sapiens*, or its replacement by other hominoids, whether nuclear wars are likely to cause an extinction, and finally what we can do to eliminate the threat of nuclear war. T-Rex, among the last of the dinosaurs on the fateful day 66 million years ago, may have wished it had a similar opportunity to change its fate.

PART I

IMPACTS, ASTEROID WINTERS, AND DINOSAURS

Chapter 2

The Power of Asteroids and Comets

When Will the Next Big One Hit?

Many different phenomena suddenly release large destructive bursts of energy on Earth. Earthquake-caused seismic waves have energy due to motions of the ground generated by the release of strain built up from drifting continental plates rubbing against each other. Explosive volcanic eruptions have energy from an expanding gas, like popping the cork off a warm champagne bottle. When the volcano fractures, it opens vents between the interior and the surface. When the vents open, dissolved gases held at high pressure in the volcanic magma are suddenly released like champagne bubbles and rapidly rush out of the mountain carrying molten rock and gases with them to high altitudes. Conventional explosives such as TNT release chemical energy. When the explosive charge is detonated, a chemical reaction occurs that generates heat. The burst of heat produces a shock wave in the air.

Asteroids and comets have kinetic energy, which is the energy of motion. The amount of kinetic energy carried by an asteroid or comet is half of their mass multiplied by the square of their speed. Since the mass is proportional to the cube of the radius, an asteroid that is 10 times larger has 1000 times as much energy as a smaller asteroid moving at the same speed. Many researchers have tried to compare the effects of asteroids of different sizes. However, it is the energy that matters, not the size. The Mt. Everest–sized asteroid that caused the Chicxulub crater, traveling more than 10 times faster than the fastest bullet from a rifle, about 20 km/s (45,000 miles per hour), had more energy than 100 million atomic bombs with yields typical of the current US arsenal. When an asteroid hits the ground, it stops moving, and its kinetic energy is converted into seismic wave energy, thermal energy, and the kinetic energy of the ejecta that are thrown out from the crater.

2.1 The Origin of the Energy of Asteroids and Comets

Asteroids have a wide range of sizes. Some dust from asteroids is made of sand-sized grains of rock, which make up most of the shooting stars that we see at night. Shooting stars become visible when the particles heat up and emit light as

Earth in Flames. Owen Brian Toon and Alan Robock, Oxford University Press. © Oxford University Press (2025). DOI: 10.1093/9780197799734.003.0002

they vaporize due to frictional heating from colliding with air molecules as the shooting stars plunge into the atmosphere from space. There are even smaller dust particles that drift through the atmosphere without burning up. These sand-sized particles contribute about 40 tons of rock per day to Earth. But many asteroids are much larger, some approaching the sizes of moons.

Surprisingly, the energy of asteroids and the energy of nuclear bombs based on fission both originate from the birth of our solar system. In the case of many asteroids, their masses and orbits were established at the dawn of the solar system.

The Sun was born more than 4.6 billion years ago in a frigid stellar nursery located in a spiral arm of the Milky Way Galaxy. The cloud of molecules composing the nursery, which no longer exists, was composed of gases, mainly hydrogen, and dust. Its total mass may have been 10 thousand times greater than the mass of the Sun. Within this stellar nursery, gravity formed a vast and relatively dense rotating clump of gas and dust called the solar nebula. The central region of the nebula contained more than 98% of its mass. Gravitational collapse of the central region formed the Sun and heated the interior of the nebula. Eventually the center of the Sun became so hot and so dense that nuclear fusion began converting hydrogen into helium. The remaining mass formed a rotating disk, something like Saturn's rings, from which the planets grew, as dust particles collided with each other to grow larger and larger. Close to the Sun, within the orbit of present-day Mars, dust particles collided to form larger lumps of rock called planetesimals—baby planets. Rather quickly, relative to the age of the solar system, the terrestrial planets—Mercury, Venus, Earth, and Mars—formed as the planetesimals collided with each other and with dust. Asteroids grew near the Sun where it was too hot for water ice to condense on them. Therefore, they are mainly rocky, although many contain water chemically bound in the rocks, and the rocky planetesimals may be the source of Earth's oceans. Past the orbit of Mars, nebular temperatures were lower, and volatile gases such as water were able to condense to form ice. Comets formed where it was cold enough for water to form ice.

The giant planets—Jupiter, Saturn, Uranus, and Neptune—became so large that they were able to gravitationally sweep up large amounts of hydrogen, helium, and other gases, as well as dust. Jupiter became so massive that its gravitational forces tore apart any growing planets between it and Mars, leaving behind a debris field of rocks, the detritus of planet formation, which we call the Asteroid Belt. Further out in the solar system, where ice balls could form, the gravitational fields of Jupiter and the other large planets hurled these ice balls out beyond the orbit of Neptune to form the Kuiper belt of comets and planetesimals. Pluto is among the largest known members of the Kuiper belt. Like Jupiter with the asteroids, Neptune may have prevented the ice balls in the outer solar system from easily sticking together and forming larger objects. The

Kuiper belt is huge, extending from the orbit of Neptune to about twice as far from the Sun as Neptune, and parts extend 30 times as far as Neptune from the sun, or about 1000 times Earth's distance from the Sun. This region is so vast, and so empty, containing only a few percent of the mass of Earth, that collisions between objects are rare. However, there may be hundreds of thousands of objects more than 100 km (60 miles) in diameter in the Kuiper belt. Because we observe comets with orbits that are predicted to repeat on million-year timescales, it is thought that there is yet another belt of icy objects called the Oort cloud. The Oort cloud is much farther from the Sun than the Kuiper belt and forms a spherical shell around the solar system, thus its long-period comets can approach Earth from any direction, while those in the Kuiper belt approach from the same plane that contains the planets. Short-period comets, such as Halley's comet, which repeats its performance every 74–79 years, have been captured by Jupiter and no longer return to the Kuiper belt or the Oort cloud. Some short-period comets have lost so much of their volatiles by passing near the Sun and forming tails that they now appear as rocky objects instead of icy ones and can be confused with asteroids. There are likely large numbers of very large comets in the Oort cloud and the Kuiper belt that could suddenly enter the inner solar system, but our observations of these are poor because they are so far away.

The mass of asteroids and comets was determined by how large they grew by collisions or condensation of volatile gases. Some asteroids are fragments of larger objects and formed by collisions between larger objects in the asteroid belt. The velocity of asteroids and comets is caused by falling toward the sun and toward the Earth as they hit it. No asteroid or comet will hit Earth with a velocity less than 11 km/s (24,600 mph), because this is the velocity of an object starting with no initial velocity falling onto Earth due to Earth's gravity. Duncan Steel (1998) found from 116 asteroids whose orbits cross the orbit of Earth that the mean velocity of impact on Earth is 17.7 km/s (39,600 mph), with a range between 12.6 and 40.7 km/s (28,000 and 91,000 mph). The speed of asteroids is dominated by the slower values in this range, with 75% being under 20 km/s (45,000 mph). However, observations of asteroid velocities show that larger objects travel faster than smaller ones, with velocities of 3–7 km diameter objects being in the range from 24 to 31 km/s (Harris and Hughes 1994). Asteroids generally travel in the same plane as the Earth and planets, so they collide like cars traveling in the same direction with a relatively small difference in velocity. Comets strike Earth at much higher velocities than asteroids, mainly because they can hit Earth from all directions, so their collision with Earth can be like cars hitting each other head on. A wide range of velocities is possible for comet collisions. Short-period comets hit Earth with velocities near 38 km/s (85,000 mph). Considering all comets observed by Steel, the mean speed of impact is 56 km/s (125,000 mph), with only 25% traveling slower than

45 km/s (101,000 mph) and 75% being slower than 66 km/s (148,000 mph). For reference, the muzzle velocity of an M16 military rifle with an M855A1 round is 0.98 km/s (2,200 mph). The fastest bullet, a 0.22 Swift, can travel about 1.3 km/s. So, a typical asteroid is moving about 20 times faster than these bullets, and a comet about 50 times faster.

Thus, the energy of colliding asteroids and comets was set during the earliest history of the solar system when their masses were determined by the collisions of dust grains and ices and their velocities were fixed by their orbits. Many asteroids were formed later in history when larger ones ran into each other and shattered.

2.2 The Probability of Asteroid and Comet Impacts

We know for the extinction of the dinosaurs that it only takes one collision by a 10 km diameter asteroid to cause a mass extinction. The effects on the environment depend on the energy released rather than the size of the asteroid or comet. These energies can be so large that we use special units and ways of writing numbers to describe them, as discussed in Sidebars 2.1 and 2.2.

Sidebar 2.1: Ways of Describing Large Energies, Conventional Energy Units, and Metric Units

Large energies are often described in units of the equivalent mass of conventional explosive needed to produce the energy.

A ton is a measure of the energy released by the explosion of a ton of the "conventional" explosive trinitrotoluene (TNT). It is not as simple as it might seem to make use of the ton unit. Ton is ambiguous because there are short tons, long tons, and metric tons, but they are all about the same. In science, we use the metric ton, sometimes spelled tonne, which is 1,000 kg or about 2,200 pounds. A metric ton is about the weight of 16 people. Further, the energy released by a ton of TNT is somewhat variable depending on how the explosion occurs. To resolve these ambiguities, a ton of explosive energy is defined to be the same as a million calories, or 4.184 million joules. A kiloton (kt) is 1,000 tons, a megaton (Mt) is a million tons, and a gigaton (Gt) is a billion tons. Calories and joules are standard energy units.

The joule (J) is a standard metric scientific unit of energy. Since a watt is a joule per second (J/s), a light bulb that uses 100 watts is consuming 100 J of energy every second. Another common energy unit is the calorie. A calorie is the amount of energy needed to heat 1 gram of water by 1 degree Celsius. A calorie equals 4.184 J. We often experience the "large calorie" unit in terms of the amount of energy released by the body when it breaks down food. A large calorie is 1,000 calories and is the energy it takes to heat 1 kg of water by 1 degree Celsius.

As scientists, we use the metric system, which is officially called using the French *Système international d'unités*, or the SI system. These units are simpler than American units because they just use multiples of 10. In most of the book, we give the equivalence between SI units and the units most Americans use and understand. But in this chapter and the next, there are many SI units, so here we give you a few equivalents all at once. One mile is about 1.6 kilometers (km). Ten feet is 3 meters (m). One inch is 2.54 centimeters (cm) or 2,540 millimeters (mm). A human hair has a diameter of about 100 micrometers (μm). One thousand miles per hour is 0.45 kilometers per second (km/s). A kilogram (kg), which is 2.2 pounds, is 1,000 grams. Freezing of water in Celsius degrees occurs at 0°C and boiling at 100°C. Add 273 to degrees C to get degrees in Kelvin (K).

Sidebar 2.2: Ways to Describe Really Large and Really Small Numbers

Large numbers can have names such as one thousand, a million, or a billion. They can also have abbreviations such as kilo, mega, of giga. However, after a certain point, we run out of names and abbreviations. In that case, we can use a notation based on powers of ten. For example, the number 10^x means a 1 followed by x zeros, and 10^{-x} means $1/10^x$. So, 10^3 is 1,000, 10^6 is 1,000,000. 10^0 is 1. Going to small numbers, 10^{-1} is 0.1, and 10^{-2} is 0.01. Powers of ten grow very rapidly. A googol is 10^{100}. The number of atoms in the universe is thought to be less than a googol, so there is not enough ink to ever write this number out as atom-sized dots on paper. If you plan to think big, you have to find a manageable way to discuss things.

In Tables 7.2 and 7.3, we estimate that about 900 Mt of energy might be released in a nuclear war with all the currently deployed weapons in the United States, Russia, France, and the United Kingdom, plus the total weapons in China. However, the energy released in the Chicxulub impactor when it hit the sea covering what is now the Yucatan Peninsula of Mexico was about 10^8 Mt, that is 100,000,000 Mt. This energy release in the asteroid collision is equivalent to 333 million nuclear weapons exploding with an average yield of 300 kt, or alternatively a 1 Mt energy release for each 5 square kilometers of Earth's surface. Fortunately, such energy releases are far beyond those possible with current nuclear arsenals. In fact, the energy in the currently deployed weapons is comparable to that in an impact by an asteroid with a diameter of a few hundred meters.

Knowing the numbers of asteroids or comets on orbits that bring them close to Earth allows us to predict the chances of being hit by an object big enough to be damaging. Near-Earth objects (NEOs) are on orbits whose closest approach to the Sun brings them within 1.3 times the distance from the Sun to Earth. More than 99% of NEOs are asteroids, NEAs. Figure 2.1 illustrates our understanding of the numbers of NEAs larger than a given diameter, updated from Harris and Chodas (2021).

Reading along the bottom horizontal scale in Figure 2.1 we find the diameter of the asteroid. The observations given by the Discovered curve are based on the amount of light reflected by the asteroids (the absolute brightness), which is related to diameter. The top horizontal scale is the impact energy, which is one half the mass of the asteroid multiplied by the square of its measured velocity.

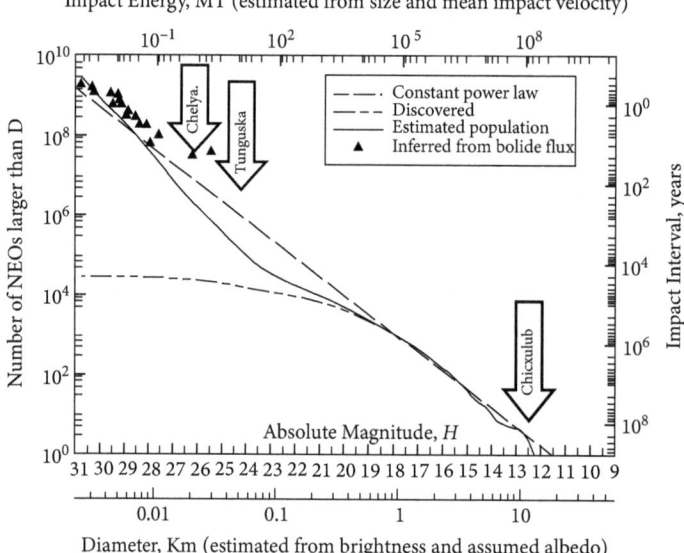

Figure 2.1 The impact energy of the asteroid is given on the top axis. The time between impacts of a given diameter is presented on the right axis. The left axis gives the number of asteroids with sizes larger than the diameter given on the bottom axis. The Discovered curve presents the number of objects whose orbits were observed before 2023. The triangles are based on objects seen or heard entering the atmosphere. The dashed curve assumes the number of asteroids larger than a given size is proportional to the diameter raised to a power. The Estimated curve is the best current best estimate for all sizes. Courtesy of Alan Harris, updated from Harris and Chodas (2021).

The vertical axis on the left is the total number of NEAs larger than a given size. There are likely about 1,000 NEAs larger than 1 km, almost 100,000 larger than 100 m, and 1 billion with energies larger than 1 kt of conventional explosives. The vertical axis on the right side is the typical time between collisions of the given size, which is the inverse of the probability of an impact in a given year by an object of the given size or larger. Several large impacts are noted in Figure 2.1, including the Chicxulub impact that killed the dinosaurs 66 million years ago, the Tunguska impact of 1908 that mysteriously left no crater in Siberia but blew down a forest, and the 2013 impact at Chelyabinsk, Russia, that injured about 1000 people.

By comparing the amount of iridium found in the debris layer left from the asteroid impact and the amount of iridium in typical asteroids, the Alvarez group estimated that the diameter of the asteroid that killed the dinosaurs was about 10 km. Once the Chicxulub crater was found, buried beneath sediments deposited over the last 66 million years and partly covered by seawater, it was also estimated that to make a crater of the size observed, about a 10 km diameter asteroid must have hit what is now the Yucatan Peninsula of Mexico.

The measurements, shown by the Discovered curve in Figure 2.1, suggest there are about two NEAs with similar size to the one that killed the dinosaurs still out there. The chance of being hit by one is about 1 in 100,000,000 in any given year. Of course, we know from observations that there are no such objects about to hit us in the near future. However, sometime in the next 100,000,000 years one would be expected to hit. Figure 2.1 indicates that in the roughly 300,000-year history of our species, *Homo sapiens*, the Earth likely has been hit, probably in the oceans, by one asteroid or comet with a diameter of 1 km and an energy near 100,000 Mt.

Of course, we have a record of some large objects hitting Earth. Martin Schmieder and David Kring (2020) describe 46 distributed debris layers and 200 craters with diameters between 1 km and 250 km that are mapped in Figure 2.2. The three largest craters—Sudbury (200 km) in Canada, Vredefort (250 km) in South Africa, and Chicxulub (180 km) in Mexico—represent about 90% of the impactor mass from these impacts. These observed craters are thought to represent only a small fraction of the impacts that actually occurred. Most objects hit the oceans, and their craters have either not been found or have been lost by subduction of the oceanic crust. Subduction occurs when continental drift forces the ocean floor beneath the continents. This process is marked by volcanic eruptions that are often explosive, such as those found in the ring of fire around the Pacific Ocean. Because of subduction, most of the ocean floor is less than 150 million years old. On the continents, smaller craters are removed by erosion. In the last 2 billion years, there may have been 20 Chicxulub- or Sudbury-sized impacts. About 27 craters formed on land during the 300,000-year history of *Homo sapiens*. About 35 craters have ages that fall in the Quaternary geologic

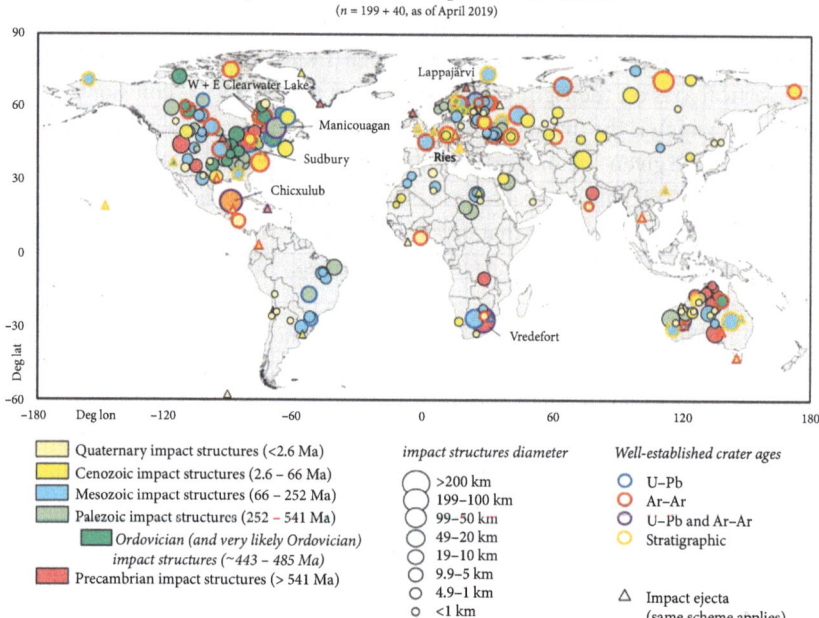

Figure 2.2 Craters are not uniformly distributed on the continents. This uneven distribution is likely an artifact of the small number of craters, continental drift, the difficulty of locating them in regions covered by forests or mountains, and the few people who have searched for them in much of Asia, South America, and tropical Africa. Courtesy of Martin Schmeider and David Kring (LPI).

period covering the past 2.6 million years. The Quaternary Period contains an ice age with numerous glacial advances and the origin of the first members of the genus *Homo*. The two largest craters in the Quaternary Period have diameters of 14 km. Such craters were probably caused by asteroids about 1 km in diameter and with energies of about 100,000 Mt. Given that the oceans have about 2.4 times the area of the land, about seven asteroids of 1 km diameter have likely hit Earth since our genus arose.

The impact of a 1 km diameter asteroid or comet is likely to have global implications. The energy of the impactor and possibly where it hits are more important than the diameter of the object. Thus a 1 km diameter object hitting the land is not 10% as important for global climate as the 10 km diameter asteroid that killed the dinosaurs, because the energy of a 1 km diameter object is only 0.1% as great as a 10 km diameter asteroid such as the Chicxulub object. There are several large craters that formed after the Chicxulub impact. Most notable are the 100 km diameter Popagai crater in Russia that is 36.63 million years old, and the 40–45 km diameter Chesapeake Bay crater in the

United States that is 34.86 million years old. These ages are suspiciously close (within millions of years) to a significant extinction event. However, the Popagai object was likely around 5–8 km in diameter, with 12–50% of the energy of the Chicxulub object, which was apparently not enough to create a major mass extinction.

Investigators such as Ahrens and Harris (1992) have argued that the annual odds of an individual dying from an asteroid collision are about the same as those of dying from an airplane crash. Most people die from diseases, and relatively few people die from accidents. This is why insurance agents love to sell accidental death insurance. About 56 million people worldwide, and about 2.8 million Americans, died in 2017 according to Table 2.1 (Ritchie and Poser 2018). However, only about 8% of deaths were from accidents in 2017. No one died from an asteroid hitting Earth in 2017, or for that matter in recorded history (though people have been injured by bouncing meteorites).

Table 2.2 lists the odds of dying from various causes as estimated by the National Safety Council (2024). One is mostly likely to die from heart disease, cancer, or an infectious disease.

However, preventable deaths (including by drugs, suicide, and accidents) do pose a significant risk of death over your lifetime. At the end of Table 2.2, we find death by bee stings, dog attacks, and lightning. People do die of these causes every year. The death rate of people on commercial airlines flying in the United States has become very low in recent decades. Between 2010 and 2019, there were five deaths on commercial aircraft. Given the 2.6 million people in the United States who died in 2017, the odds of death in an airplane crash were 1 in 5.6 million per year or 1 in 77,000 over a typical lifetime of 72 years. This death rate is quite uncertain due to the low number of airplane deaths in recent history. With asteroid collisions, it is likely that an impact of a 10 km diameter asteroid will kill nearly everyone on the planet. However, such an event would only occur about once every 100 million years. Your chances of dying from a 10 km asteroid impact in your lifetime are then about 1 in 1.4 million. However, smaller asteroids are likely to kill a large fraction of the population. Unfortunately, we cannot precisely predict how many would die from an impact of a given size.

Table 2.1 Causes of Death in 2017 (Ritchie and Poser 2018)

Cause of Deaths	Worldwide (million)	American (million)
Non-communicable disease	41.1	2.5
Communicable, infectious, maternal, and neonatal disease	10.4	0.14
Accidents	4.5	0.19

Perhaps 50% of the population would die if a 1 km diameter asteroid hit, which has a probability of occurring of about 1 in a million every year. Therefore, the odds of dying from a 1 km asteroid impact is about 1 in 28,000. As indicated in Table 2.2, the odds of death from an asteroid collision span a large range because the odds depend on the size of the asteroid that would kill a large fraction of

Table 2.2 Lifetime Odds of Death for Selected Causes, United States, 2017 (National Safety Council 2024)

Cause of Death	Odds of Dying
Heart disease	1 in 6
Cancer	1 in 7
All preventable causes of death	1 in 25
Chronic lower respiratory disease	1 in 26
Suicide	1 in 86
Opioid overdose	1 in 98
Motor vehicle crash	1 in 106
Fall	1 in 111
Gun assault	1 in 298
Pedestrian incident	1 in 541
Motorcyclist	1 in 890
Drowning	1 in 1,121
Fire or smoke	1 in 1,399
Choking on food	1 in 2,618
Bicyclist	1 in 4,060
Sunstroke	1 in 7,770
Accidental gun discharge	1 in 9,077
Electrocution, radiation, extreme temperatures, and pressure	1 in 12,484
Sharp objects	1 in 29,483
Hot surfaces and substances	1 in 45,186
Hornet, wasp, and bee stings	1 in 53,989
Cataclysmic storm	1 in 54,699
Passenger on commercial airplane	1 in 77,777
Asteroid collision (1 km or 10 km object; based on data in Figure 2.1 and human lifetime of 72 years)	1 in 28,000— 1,400,000
Dog attack	1 in 118,776
Lightning	1 in 180,746

the population. However, they are comparable to lightning and dog attacks—and possibly more likely than aircraft accidents. As we show later, a nuclear war could kill most of the people on Earth (by starvation). We don't know the odds of a nuclear war occurring, but ridding the world of nuclear weapons will prevent this.

These statistical comparisons are interesting. In the case of asteroid collisions, the odds of a collision with a large asteroid are low, but the consequences might be to kill a large fraction of the Earth's population. Dog attacks, lightning strikes, and airplane crashes do kill people regularly, but the number of people affected is very small. It would take a lot of plane crashes, about 140 million at the current rate, to kill the population of Earth.

One surprise on Figure 2.1 is found at the short end of the timescale. Every year an object slightly larger than 2 m in diameter and with an energy of about 2 kt hits Earth. These objects are observed by various satellites designed to detect nuclear explosions, and they are sometimes seen by people at the surface. The bright flash as the object enters the atmosphere and disintegrates can be seen by infrared sensing instruments, and the sound waves from the object can be heard by infrasound sensors, which detect low-frequency sound waves. Every few years an object enters with the energy of the atomic bomb used on Hiroshima, Japan, in World War II (about 15 kt).

Why don't we live in fear of a sudden devastating explosion at the ground? The answer to this question came from decades of research on the Tunguska event in Siberia. The event occurred on swampy forest land near the Podkamennaya Tunguska River on the morning of June 30, 1908. A man sitting on a porch at a trading post about 60 km away from the impact site was knocked off his chair, and he felt an intense heat pulse. The impact site was not investigated until nearly 20 years later. About 2,000 km^2 of forest was knocked down in a radial pattern shaped like a butterfly, but there was no crater on the ground and no evidence of asteroid debris. The event was not just local. In Europe, the sky was so bright the night after the impact that people could read newspapers outside, and pressure waves were detected as far away as Washington, DC, and Indonesia.

Due to the lack of a crater or asteroid debris, many theories were advanced to explain the event, ranging from black holes passing through Earth to alien spaceships exploding. However, we now know that a stony asteroid or possibly a comet fragment hit Earth. The ram pressure on the leading edge of the object as it passed through the atmosphere created a shock wave propagating through the asteroid (Chyba et al. 1993). Stony asteroids and comets do not have much mechanical strength, and the shock wave blew the object into tiny pieces. Iron asteroids are strong enough to not disintegrate, which is why iron meteorites are commonly found on the ground. However, stony meteorites are seldom found even though they are far more common than iron asteroids in space. Similar events in which the impactor disintegrates in the atmosphere have been observed many times and are called meteors, fireballs, or bolides. Estimates of the energy and size of the Tunguska object are uncertain due to a trade-off between explosion altitude and energy and other factors. Estimates of its diameter range from about 50 to 200 m, comparable to the size of a football field. Estimates of the energy release range from 3 Mt to around 30 Mt. The lower energy release assumes the asteroid disintegrated only a few kilometers above the surface, while the large value requires disintegration above 10 km.

Another dramatic event occurred on February 15, 2013, in Chelyabinsk, a city with a population greater than 1 million located just east of the Ural Mountains in Russia. This event was recorded on cameras by many people in the city as well as by satellites. The meteor exploded at an altitude of 25–30 km and generated a shock wave that injured over 1,000 people, mostly by broken glass. More than 7000 buildings were damaged by the shock wave in six cities over an area around 100 km wide and tens of kilometers long. Infrasound waves were observed going around the Earth twice.

The object produced a large luminous trail due to friction with the air as it passed through the atmosphere on a path nearly parallel to the ground. People rushed to their windows to see what was happening. Then the shock wave hit and shattered window glass onto them. Some people who were outside at the time received burns from the heat radiated by the hot luminous trail. This object is thought to have been about 20 m in diameter, traveling at a speed of about 20 km/s, and had an energy estimated in the range from 470 to 590 kt when it hit Earth (Popova et al. 2013). Only about 0.05% of the original asteroid is thought to have hit the ground. About 76% was vaporized by the heat generated by friction with the atmosphere, and the remainder was pulverized to dust by the shock wave. Fortunately, the immense energy release was far above the surface; otherwise, the city might have been reduced to rubble. This high altitude for the energy release was partly due to its passage at a very shallow angle, rather than descending to the ground at a steep angle. Small pieces of the object were found, and it was an ordinary chondrite, the most common type of stony meteoritic material to hit Earth. It may have once been part of a larger LL-chondrite asteroid that formed before the Earth-moon system and that underwent collisions with other asteroids through time (Kring and Boslough 2014).

The events at Tunguska and Chelyabinsk show that most small asteroids disintegrate in the atmosphere well above the surface. Their shock waves have lost much of their destructive potential by the time they reach the ground. Figure 2.3 shows predictions of the altitude at which various compositions of asteroids (stones, carbonaceous, and iron) and long- and short-period comets entering the atmosphere at 45 degrees would disintegrate as a function of their energy. These smaller objects are suddenly slowed by drag when they encounter a region

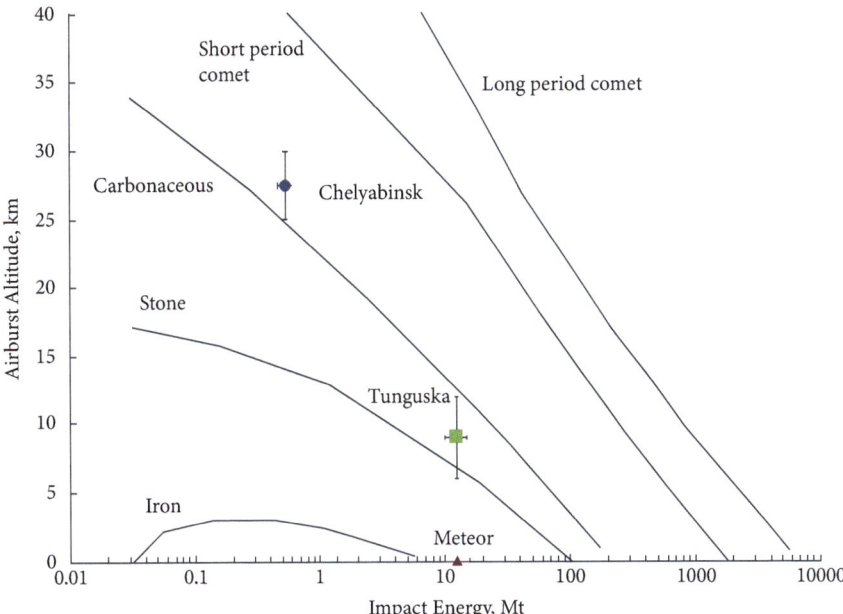

Figure 2.3 Predicted airburst altitudes for comets and asteroids incident at an angle of 45° relative to the ground. The estimated energies and disintegration altitudes for the object that formed the Meteor Crater in Arizona, the object that caused the Tunguska, Russia, event of 1908, and the asteroid that hit Chelyabinsk, Russia, in 2013 are plotted for comparison. The top two curves are for comets. Modified from Toon et al. (1997).

© 1997 American Geophysical Union

of the atmosphere with a high enough air density. The sudden deceleration produces a strong shock wave within the asteroid or comet that blows it into pieces and may vaporize parts of it. Stones and carbonaceous asteroids may be rubble piles left from the collisions of smaller rocky objects that are held together by gravity. Comets are rather weak since they are made of ice that might be fractured, and even quite energetic ones might blow apart before hitting the surface. In contrast, strong iron asteroids with a wide range of energies will strike the surface. Meteor Crater in Arizona, which has a diameter near 1 km, was made by an iron object with an energy near 10 Mt.

Figure 2.1 illustrates a number of additional interesting points. First, there are around 1 million NEOs as large or larger than the Tunguska impactor. Objects the size of the Tunguska impactor likely hit Earth every few hundred years, but they are too small to see in space unless they are close to Earth or large telescopes are used. Tunguska devastated an area of about 2000 km², about five times greater than the area of Denver, Colorado, and slightly larger than the area of London. Had Tunguska hit a city, major loss of life would have occurred.

Any casual follower of the news will know that asteroids pass by Earth very frequently at distances about as far away as the moon. In fact, the day the city of Chelyabinsk was struck, another slightly larger asteroid, which was predicted in advance and widely publicized, passed by Earth. Its orbit was different from that of the Chelyabinsk impactor, suggesting the two objects were not related. Table 2.3 lists the frequency of an Earth impact and the frequency of passage between Earth and the moon for various-sized objects. In the popular press, passing between the Earth and the moon, which are about 380,000 km (240,000 miles) apart, is apparently a near miss and worthy of announcing. This is like saying "look out!" because someone threw a baseball in Los Angeles and they might break a window in New York. An object the size of the asteroid that killed the dinosaurs passes nearby every 30,000 years. However, an object with a diameter near 140 m and an energy of 100 Mt, about twice as energetic as the largest nuclear weapon ever exploded, the Tsar Bomba, passes between Earth and the moon every 6 months. Of course, just crossing Earth's orbit does not mean a collision will occur.

In 1993, Brian attended a meeting in Tucson, Arizona, whose goal was to determine which sized comets and asteroids we should try to stop from hitting Earth and how we might keep them from hitting Earth. The Cold War had just ended, and it was thought peace had broken out on Earth, that nuclear weapons would be abolished, and that countries would solve their differences by negotiation and cooperation. This left the nuclear weapons experts with nothing to do. As a consequence, this meeting not only had theoretical astronomers such as Carl Sagan but also impact luminaries, such as Dave Morrison and Clark

Table 2.3 Impact frequency, the cumulative frequency for objects larger than the given diameter. The diameter and energy are related assuming a density of 2,500 kg m^{-3} and a velocity of 15 km s^{-1}.

Impact Energy (Mt)	Diameter (km)	Impact Interval (years)	Interval within Lunar Orbit (years)
10^{-2}	0.007	0.5	Every year
10^{-1}	0.014	5	Every year
10^{0}	0.03	30	Every year
10^{1}	0.07	200	Every year
10^{2}	0.14	2000	0.5
10^{3}	0.3	1×10^{4}	3
10^{4}	0.7	6×10^{4}	20
10^{5}	1.4	3×10^{5}	90
10^{6}	3	2×10^{6}	600
10^{7}	7	1×10^{7}	4,000
10^{8}	14	9×10^{7}	30,000

Chapman, who had spent much time studying impact probabilities and consequences. The meeting also included observational astronomers such as Eleanor
Helin, Carolyn Shoemaker, and Gene Shoemaker developing plans to find hazardous asteroids, and geologists such as Jan Smit, who had studied the impact
debris left from the Chicxulub impact, as well as Russian and American nuclear
weapons experts such as Edward Teller. In a spirit of cooperation, the nuclear
experts from America agreed to share their previously classified data on detections of asteroids from the sounds of their impacts on the atmosphere with the
Russians, who were thought to be better able to analyze them and determine the
frequency and size of the currently occurring impacts. Current analyses of the
detections of small asteroid are illustrated in Figure 2.4.

We can detect the energy radiated by meteors as they pass through the atmosphere and disintegrate due to the shock wave. While some data have been
obtained by lightning sensors on satellites in the past few years, most data were
taken by satellites trying to detect nuclear explosions. Figure 2.4 illustrates data
over 36 years (Jet Propulsion Laboratory 2024). Yellow and red colors represent
impacts with an energy greater than 1 kt. During the period of observation, 96
objects with energies above 1 kt were observed, and 14 had energies above 10
kt, similar to the Hiroshima atomic bomb, whose energy was near 15 kt. Note
that a log scale gives the power of ten of the energy in kilotons. For instance, 2
indicates 100 kt, and 0 indicates 1 kt.

An interesting debate in the Arizona meeting broke out about a graph like
Figure 2.1. In the meeting, the nuclear weapons experts proposed a variety of
ideas, including putting nuclear weapons on the moon to use to divert asteroids
without having to have such weapons on Earth any longer. Also, the nuclear

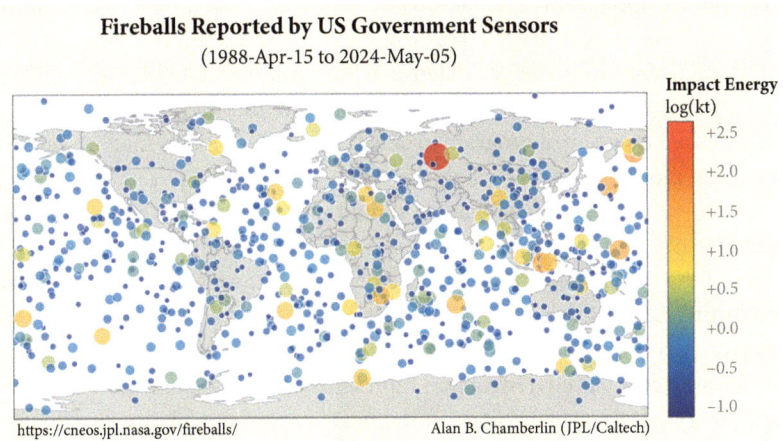

Figure 2.4 Fireballs reported by US government sensors between April 1988
and May 2024 (Jet Propulsion Laboratory 2024)

weapons experts suggested that they should destroy any small asteroid coming between the Earth and the moon with nuclear weapons to practice averting a collision. Carl Sagan objected that this was merely a ruse to continue to keep nuclear weapons and that it might be possible for some demented person to divert an asteroid to hit the Earth (Sagan and Ostro 1994).

In contrast, the astronomers noted that small objects would disintegrate in the atmosphere and did not need to be blown up. Astronomers, who did not want to spend precious telescope time tracking a million small asteroids and did not want nuclear weapons on the moon, countered that only the few thousand objects larger than 1 km in diameter should be of concern.

It was argued that in the past natural calamities on Earth had often killed millions of people, and that the Second World War led to around 50 million deaths. Therefore, we should ignore asteroids and comets that might destroy a city and only be concerned with those that might cause global problems with fatality levels exceeding those of the Second World War. A cost-benefit approach was described by Ahrens and Harris (1992), who suggested that the odds of an individual on Earth dying from a Tunguska-sized impact is about 1 in 300 million per year, while the chance of an individual's death from a several-kilometer-sized impact is about 1 in 2 million per year, which was then about the odds of death in an airplane crash. The odds of death from a several-km sized impact are about 100 times higher than from a Tunguska-sized impact. A Tunguska-sized impact occurs 1000 times more frequently than a km-sized impact. However, the km-sized impact could affect a large fraction of the Earth, while the city killer-sized Tunguska impact would affect an area that is one hundred thousand times smaller. Given the odds of death from a km-sized impact, they argued that we should fund asteroid protection at about the same rate as protection from aircraft crashes, which they estimated to be $10 million to $100 million per year.

The debate about which dangerous objects need to be prevented from hitting Earth has not been settled. However, in 2019 and 2020, NASA spent $150 million each year on planetary defense, which included satellites to visit large asteroids and investigate how to prevent them from hitting Earth.

NASA does now have a program to detect dangerous objects that has successfully found most of the expected objects larger than 1 km (Figure 2.1). However, the size of the smallest object that we should protect ourselves against remains uncertain.

Part of the uncertainty about the size of the object we should try to prevent from hitting Earth is due to the level of damage that society chooses to accept. However, part of the uncertainty is due to the difficulty in estimating the damage from an impact. Assured destruction represents the type of damage that we can calculate accurately, as we discuss in the next chapter along with clues to what did cause the extinction of the dinosaurs.

Chapter 3

Clues from Craters, Assured Destruction, and Ejecta Layers

"It's really hot," Brian thought as he meandered his way down the rocky slope of the Meteor Crater in Arizona on a summer day in 1983. He was on a trip led by the legendary Gene Shoemaker to better understand how craters form by impacts and nuclear explosions as part of a panel organized by the US National Academy of Sciences to determine if nuclear war might trigger a nuclear winter. Gene was famous for having proved the Meteor Crater was created by an iron meteorite hitting Earth and was not a product of a volcanic explosion. Gene trained the lunar astronauts on crater geology using Meteor Crater as an example as they prepared to leave Earth for the first time to explore the moon. While Brian and the other sweating members of the panel gingerly made their way down the boulder-strewn slope, the ever-ebullient Shoemaker leapt from boulder to boulder well ahead of them. Finally, he stopped to lecture about the distinguishing characteristics of Meteor Crater, probably the best preserved crater on Earth.

It is a simple crater, with a circular shape, squared off by preexisting faults, about 1.2 km in diameter as shown in Figure 3.1. The somewhat eroded rim of the crater stands about 40 m (150 ft) above the surrounding plain, and the crater is about 170 m (550 ft) deep. The crater floor has more than 200 m of rubble and post-impact lake sediments above the bedrock. The crater was formed when a nickel-iron asteroid plunged to the ground nearly 50,000 years ago. The size and velocity of the object are still debated. However, it was in the range about 30 to 60 m (90 to 180 ft) in diameter and traveling with an energy in the vicinity of 10 Mt when it hit the ground. Miners tried to dig up the meteorite for its ore, which they thought would be in the center of the crater due to its nearly circular shape, but in fact the major remnants are buried under part of the rim. The asteroid may have partially disintegrated during its motion through the air.

As Gene talked, Brian looked down toward his feet and discovered a small piece of the meteorite, an iron rock with little splashes of nickel, shaped something like a dried-up apple core (Figure 3.2). It was not an important find; the region around the crater is littered with fine debris. But as he held it, Brian thought, "Wow, I'm holding something that may be older than the Earth itself!" Iron meteorites are believed to originate from the cores of asteroids with

Earth in Flames. Owen Brian Toon and Alan Robock, Oxford University Press. © Oxford University Press (2025). DOI: 10.1093/9780197799734.003.0003

Figure 3.1 Meteor Crater in Arizona, United States. From NASA Earth Observatory

diameters between a few kilometers and 100 kilometers that differentiated by melting either due to radioactive heating or impacts of planetesimals during the early formation of the solar system. This meteorite in Figure 3.2 is a small reminder of our planet's ultimate origins.

Asteroids and comets passing through the atmosphere create shock waves and a bright light due to heating by friction with the air. The light is capable of igniting fires on the ground. Most asteroids with energies larger than 100 Mt and diameters above 100 m, as well as comets with energies greater than 1000 Mt or larger than 300 m in diameter, will hit Earth's surface instead of blowing up in the atmosphere, as shown for smaller objects in Figure 2.2. The fireball rising from the impact site will radiate energy that may start fires and cause flash burns in addition to those caused by the hot object moving through the air. Impacts may deposit toxic compounds from vaporization of the target as well as the asteroid or comet when they hit the ground. The shock waves in the ground propagate as earthquakes, and those in the ocean create tsunamis. Shock waves and tsunamis can devastate parts of Earth but are not likely to cause mass extinction events since they don't involve the entire planet. The impact energy is eventually converted into heat. Impacts of asteroids and comets heat the air and ground locally to very high temperatures, which increase with the size of the object. Early in Earth's history, it is thought that some objects hundreds of kilometers in diameter struck the planet and released so much energy that they boiled the upper

Figure 3.2 Left: A small piece of the nickel-iron meteorite that made the Meteor Crater in Arizona found by Brian. Right: A carbonaceous chondrite of type CM2, similar to the meteorite that is thought to have created the Chicxulub crater. This meteorite is part of the Murchison meteorite on display at the US National Museum of Natural History in Washington, DC. Such meteorites may contain amino acids and grains of separate minerals that predate the solar system. It has been suggested that some of the chondrules (the white and yellow dots in the figure) in the Murchison meteorite are 7 billion years old, more that 2 billion years older than the solar system.

portions of the oceans. Possibly, these large impacts, which were more common early in Earth's history, destroyed the first life to arise, explaining why it took more than 500 million years for life to gain a foothold and create the earliest fossils we have discovered. Such large objects still visit the inner solar system. The recently discovered comet Bernardinelli-Bernstein is thought to be more than 120 km in diameter and is expected to come closest to the Sun and the Earth in 2031. Fortunately, it will remain outside the orbit of Saturn, so it can't hit Earth on this visit.

One of the mysteries of the K-Pg extinction event is why the impact caused a global catastrophe. (K is an abbreviation for *Kreide* ("chalk" in English), the German word for the Cretaceous geologic period, and Pg is the abbreviation for the Paleogene geologic period. These two geologic periods were separated by the asteroid impact about 66 million years ago that killed most of the dinosaurs.) The global consequences of impacts include climate change and loss of sunlight from debris injected into the atmosphere by the impact, fires started by the re-entry of the impact debris, and possibly acidification of the oceans from impact debris falling into the oceans. Each of these phenomena is difficult to predict and evaluate for its effects on the biota. However, like nuclear explosions, impacts produce blast waves in the atmosphere, earthquake waves, and tsunamis. These effects almost certainly will occur, and the damage from them should be possible to predict, so they are called assured destruction.

3.1 Craters

Not all impacts will produce craters. As discussed in Chapter 2, stony asteroids smaller than about 100 m in diameter and comets smaller than about 300 m in diameter are so mechanically weak that they disintegrate in the atmosphere. Larger, or stronger, objects will hit the surface and create a crater.

Crater formation is primarily dependent on the energy of the impactor or bomb. Even impactors 100 m in diameter, about the length of an American football field, have much more energy than the largest nuclear weapon ever exploded, the Tsar bomb with an energy of 50–58 Mt. Figure 3.3 shows the kinetic energy of comets and asteroids of given diameters (Shoemaker et al. 1990). A 100 m diameter asteroid has an energy of about 100 Mt. Due to their higher velocity, comets of a given diameter usually have higher energy than asteroids of the same diameter despite their lower densities.

Figure 3.3 also links the size of the crater produced by an impact or a nuclear ground burst explosion to the size of the crater that is produced (Shoemaker et al. 1990). Figure 3.3 shows that a 1 Mt surface nuclear blast will yield a crater about 400 m in diameter. For low energies, crater formation is relatively simple. These craters, such as Meteor Crater, are bowl shaped. A 1 Mt nuclear weapon crater would be about 100 m deep. For a 1 Mt ground burst, the pressure generated by the explosion is huge, like the pressure at the center of the Earth. This sudden impulse produces a shock wave in the ground. The extreme conditions of temperature and shock pressure melt parts of the ground and fragment parts of it. These materials are then hurled from the crater with tremendous force to form an ejecta curtain, most of which falls within a few radii of the crater. For small energy releases, such as a 1 Mt nuclear explosion, the crater formed just after the blast is the same as the persistent crater. However, for larger energy releases, the crater changes shape after it is formed in response to geologic forces.

Figure 3.3 links the observed size of the Chicxulub crater, formed by the impact that killed the dinosaurs, to the size of the asteroid and its energy. The crater, now buried about a kilometer deep under sediments laid down in the sea after the impact occurred, is observed to be between 180 and 200 km in diameter.

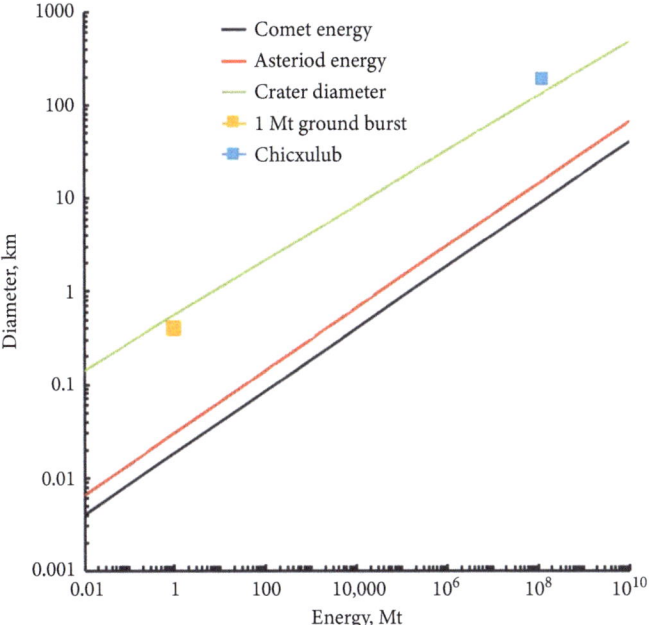

Figure 3.3 The kinetic energy of asteroids (red line) and comets (black line) as a function of their diameter. For this figure, it is assumed that comets have a density of 1000 kg/m^3 and a speed of 50 km/s. Asteroids are assumed to have a density of 2500 kg/m^3 and a speed of 20 km/s. Also shown is the expected diameter of the crater for either impacts or nuclear ground bursts of a given energy (green line). It is assumed that the density of the target material is the same as that of the impactor and that the impactor hits from a direction that is perpendicular to the ground, but the results are not very sensitive to these assumptions. The yellow square on the graph indicates the predicted crater diameter produced by a 1 Mt nuclear ground burst, and the blue square is the observed crater diameter of the crater produced by the Chicxulub impact that created the K-Pg extinction event. Modified from Toon et al. (1997).

The current best estimate is that the asteroid impacted with an energy of 120 million Mt, probably at a slightly oblique angle between 45 and 60 degrees. If it were traveling at 18 km s^{-1}, a typical speed for small objects entering Earth's atmosphere, its diameter would have been in the range from 12.2 to 14.4 km depending on the angle (Morgan et al. 2022). If its speed were higher, its diameter would have been smaller to produce the same energy. Observations show that larger asteroids travel faster than smaller ones. Asteroids with diameters from 3 to 7 km would hit Earth at speeds between 24 and 31 km s^{-1}, with the most likely speed being about 28 km s^{-1} according to observations (Harris and Hughes 1994). With this higher speed, the expected diameter of the Chicxulub asteroid is between 9.1 and 10.7 km.

Considering that the yields of the largest nuclear weapons in the current US and Russian arsenals are about 1 Mt, the Chicxulub impact is equivalent to a nuclear conflict with 100 million high-yield nuclear weapons, far more than the approximately 13,000 weapons now in existence. Not all the energy of the Chicxulub impact was distributed globally, but if 20% of it were, it would be equivalent to having a 1 Mt nuclear weapon explode about every 5 km over the surface of the Earth.

For large impactors, crater formation is complicated, as illustrated in Figure 3.4. The eventual crater size depends on the relaxation of the ground after the impact. Generally, these larger craters, mostly observed on the moon and other planets, have flat bottoms due to infilling by molten and pulverized rock. There is an outer elevated crater rim and either a central peak or elevated rings depending on the crater size. The Chicxulub crater has multiple rings. At the time of the impact, a transient crater formed that was 80–100 km in diameter (Morgan et al. 2022). The ejecta blown out of the crater removed weight from the lower rocks, which had been shoved downward by the force of the impact. These lower rocks were intensely shocked, which probably caused them to behave like a viscous fluid, or the shocks might have created impact-generated faults. The lower rocks respond by moving upward to form a central peak in mid-sized craters. However, in very large craters such as Chicxulub, the central peak cannot support its weight, so it sags downward and generates a series of rings around it. In the case of Chicxulub, melted rock and debris-filled water flowed back into the crater, partially filling it as the debris settled (Morgan et al. 2022).

3.2 Blast Waves, Earthquakes, and Tsunamis

The formation of the Chicxulub crater led to strong earthquakes, tsunamis, and an atmospheric blast wave. Each of these waves propagated across Earth with a different speed, amplitude, and share of the original energy. Most of the impact energy, like most of the energy in a nuclear blast, goes into thermal energy or kinetic energy of the impact debris.

While poorly known, it is thought that about 3% of the impact energy of a large impact at the surface will go into atmospheric shock waves. This is a lower fraction than the blast waves from a nuclear explosion, because these large impacts blow the atmosphere above the impact site into space and so there is no air to carry the shock. With this energy transfer fraction, the area within the

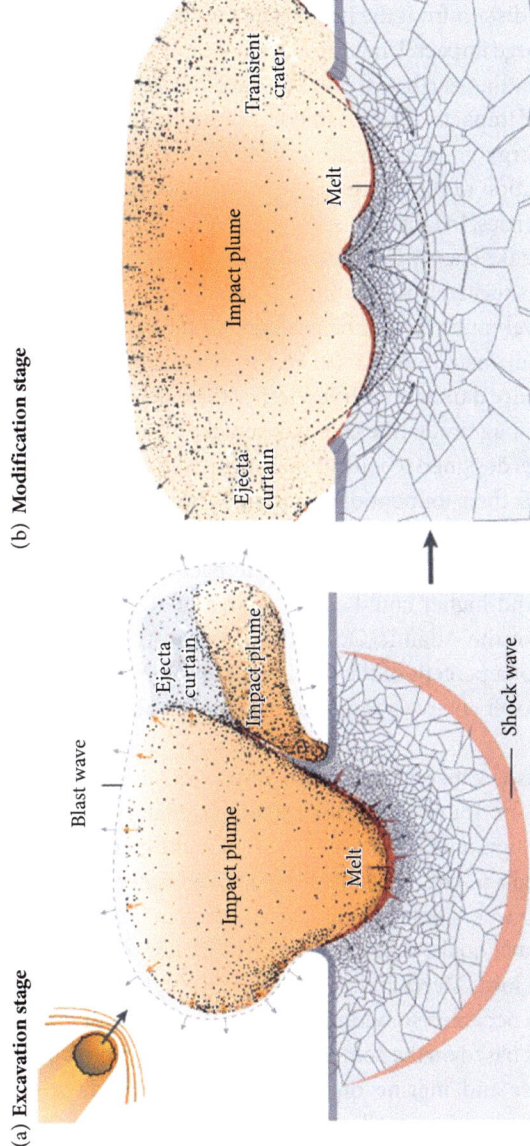

Figure 3.4 Stages of complex crater formation. a) In the excavation stage, rocks at the impact site are displaced to form a cavity with an uplifted rim and are ejected within a plume and ejecta curtain. b) In the modification stage, the cavity rim collapses downward and inward, whereas rocks in the central area initially collapse inward and upward and then downward and outward (see arrows). The size of the final crater increases with increasing impact energy, and its morphology changes from a central peak crater to a peak ring crater, and finally a multi-ring basin for craters as large as the one at Chicxulub. Drawing from Morgan et al. (2022).

overpressure level where brick buildings might be destroyed is about 10 million km^2 (3.8 million square miles), which is about the area of the United States, or about 2% of the area of Earth (Toon et al. 1997).

Crater formation produces strong shock waves in the ground near the crater. However, earthquakes far distant from the impact site only contain a small fraction of the energy from a large impact. Underground nuclear tests produce about 5% of their yield in the leading earthquake shock wave (Glasstone and Dolan 1977), while estimates for impacts indicate one-tenth to one-thousandth of a percent of the impact energy ends up in distant earthquake waves. Much of the uncertainty in the fraction of the energy in far-distant earthquake waves is related to the crustal properties. In locations such as California, the crust is heavily faulted, and the earthquake energy is absorbed quickly in the broken rock. In other locations, the crust is relatively unfractured by previous earthquakes, and the seismic waves can travel much farther before they are absorbed. Very large earthquakes measure about magnitude IX (9) in the modified Mercalli scale, which is designed to measure damage observed rather than shock energy measured by a seismometer. A magnitude IX earthquake cracks the ground, damage is considerable in buildings designed to withstand earthquakes, houses begin to collapse and are shifted off their foundations, and pipes break. Such an earthquake is roughly equivalent to a Richter scale magnitude 7, which is based on wave energy. If the seismic wave energy is 1% of the impact energy, then earthquakes of magnitude IX and higher could cover an area of about 100 million km^2, similar to the area of the Atlantic Ocean. However, if the distant wave energy is one-hundredth of a percent of the impact energy, then waves exceeding magnitude IX would cover an area about 20 times smaller, about half that of the United States.

Recently, remains of fish whose gills contain impact-generated spherules (small spheres of rock) were found in North Dakota (DePalma et al. 2019). The evidence suggests a magnitude X–XI (10–11) earthquake generated by the Chicxulub impact triggered a seiche, a seismically generated water wave, at least 10 m high in a river. Similar waves were observed in Norwegian fiords, caused by the 2011 Tohoku, Japan, earthquake that was the fourth-largest recorded earthquake on Earth and had a local Mercalli magnitude of X (10). It is thought that the event in North Dakota occurred sometime between 13 minutes and 2 hours after the Chicxulub impact based on the computed arrival times of the spherules. In this deposit, freshwater and marine organisms are intermixed, probably because the wave was generated in a shallow sea some tens of kilometers from the location of the river deposits.

Tsunamis, of course, depend on an impact in the ocean or near enough to the ocean for undersea avalanches to generate a wave. There is clear evidence for large amplitude tsunamis along coastal areas of the current Gulf of

Mexico following the Chicxulub impact. The height of a tsunami depends on the depth and topography of the coastlines where they come ashore. The waves can be greatly amplified by topography. It is simpler to predict the height of the open ocean wave, which is much smaller. It is difficult to determine how far the waves travel since that also depends on whether they were stopped by running ashore. Unfortunately, the shorelines are poorly known from 66 million years ago. However, the largest twentieth-century tsunami was generated by a 1960 earthquake in Chile, and that tsunami killed people in Hawaii and Japan. The 2004 Indian Ocean earthquake had a Mercalli intensity of IX off the north coast of Sumatra. Its tsunamis killed more than 200,000 people including small numbers as far away as Tanzania and South Africa. It is estimated that about 0.1% of the Chicxulub impact energy may have gone into distant tsunamis. Even in a relatively shallow 2 km deep ocean, 10 m high open-ocean waves could cover an area as large as the Pacific if they were not blocked by land.

In summary, tsunamis, earthquakes, and blast waves after the Chicxulub impact involved only a few percent of the impact energy. While the earthquakes, tsunamis, and blast waves were unprecedented in magnitude compared with events during human history, their geographic reach was limited, and they are not likely to have been a direct cause of extinction at the global scale. However, they were highly destructive in the area of the Gulf of Mexico. Earthquake waves travel at speeds near 1–10 km/s (3600–36,000 km/hr or 2200–22,000 mph) depending on the type of wave and the type of rock. Tsunami waves travel at speeds of near 0.15 km/s (500 km/hr or 300 miles/hr) in a 2 km deep ocean like the Gulf of Mexico. Due to their greater speeds, earthquakes occurred first in the Gulf of Mexico, causing submarine landslides over wide areas. Then tsunami waves reached distant shorelines and shallow water, further stirring ocean sediments and washing debris from coastlines into the water. Tsunami waves also reflected and passed back and forth across the Gulf of Mexico several times. On top of the debris swept up by the tsunamis and earthquakes, material ejected from the crater when it formed fell into the sea, creating layers outside the crater that are up to 80 m thick in what is now part of the Yucatan Peninsula. The deposits that are currently land based as well as recent cores taken in the Chicxulub crater reveal the sequence of deposition (Gulick 2019). A core taken inside the crater within the peak ring reveals on top a 111 m thick layer of sedimentary rock deposited after the impact, then a 104 m thick layer composed of angular fragments of melted rock and glass that is similar to rocks found in other impact craters, followed by a 25 m thick layer of melted rock mixed with some broken rock fragments. Finally, there is the underlying crustal material, mostly granite that has been shocked and mixed with some remnants of impact debris.

Between the sequence of layers largely composed of debris shaken loose by earthquakes as well as from tsunamis and a top layer composed of post impact sedimentary deposits, there is an iridium-rich layer (Goderis et al. 2021). Iridium is an element that is relatively rare on Earth's surface but concentrated in meteorites. The iridium in this transitional layer likely deposited within a few years of the impact and is represented worldwide by a layer about 3 mm thick, called the Distal layer. Some early studies in the Gulf of Mexico noted that the iridium layer was separated from the crater by 100 m of sediment. These studies interpreted the sediment layer as normal sedimentation in the Gulf of Mexico and announced that the crater predated the K-Pg layer found worldwide by hundreds of thousands of years. It is clear now that this interpretation is incorrect. The sediment layer is a result of tsunamis and earthquakes causing slumping on the edges of the Gulf of Mexico as well as ejecta from the crater falling into the Gulf of Mexico just after the impact. The iridium-rich layer was emplaced on top of the sediment layer within a few years as fine particles settled out of the atmosphere and water column (Bardeen et al. 2017) and deposited on top of the coarser debris left from the impact.

3.3 Ejecta Layers

The assured destruction phenomena associated with the Chicxulub impact, such as tsunamis, shock waves, and earthquakes, were not likely to have caused global extinctions because of their limitations to small areas of the Earth. Therefore, we must look at the ejecta from the crater for things that might have a global impact.

The material thrown out of the crater, the ejecta shown in Figure 3.4, fell back to Earth and formed a thin layer of material across the planet. Part of the layer is illustrated in Figure 3.5. This layer is now called the K-Pg layer. Walter Alvarez (1997), a geologist studying the Apennine Mountains, became interested in the K-Pg layer as part of a project to date the various geologic layers near Gubbio, Italy, as told in his fascinating book *T-Rex and the Crater of Doom*. Walter and his father, Luis Alvarez, pictured in Figure 3.5 at Gubbio, thought about how to determine the length of time needed to create the K-Pg layer. They decided to do this by measuring the amount of meteoritic material in the layer, because such material has a relatively constant rate of deposition everywhere on Earth, about 40 tons per day spread over the entire planet. To their surprise, the layer contained a very high concentration of the element iridium, which is a sign of the presence of meteoritic material. In fact, they concluded the layer was a remnant of an impact by a single object with a diameter of about 10 km (Alvarez et al. 1980).

Figure 3.5 Left: Brian points to an exposure of the K-Pg intermediate layer at Trinidad Lake State Park in Colorado, United States. The layer is the white stripe, which is sandwiched between dark coal layers above and below. Just at the top of the intermediate layer, which is several centimeters thick, is a several millimeters thick Distal layer that is rich in iridium. This is one of the best exposures in the world. You can touch the remnants of the asteroid that killed the dinosaurs with just a short walk! Right: Louis and Walter Alvarez at the site of the Distal layer (under Walter's fingers) at Gubbio, Italy.

There are four basic ejecta from the crater: crushed rocks, melted rocks, vaporized rocks, and gases released from the rocks. In the case of Chicxulub, there was also seawater ejected from the ocean above the impact site, which might have been 500 m deep. The amount of crushed, melted, and vaporized ejecta varies by about a factor of two over the range of likely asteroid speeds. Simple estimates suggest that the mass of crushed rock is in the range from 60 to 315 times the mass of the impactor. The mass of melted rock is in the range from five to 15 times the mass of the impactor. While the mass of vaporized rock ranges from zero to two times the mass of the impactor. If we assume the diameter of the Chicxulub asteroid was 12 km, the mass of the asteroid would be about 2.2×10^{15} kg. For reference, the mass of Mt. Everest is estimated to be about the same as the Chicxulub asteroid. Therefore, the amount of material hurled over the Earth, mostly as crushed rock, was hundreds of times larger than the mass of Mt. Everest.

Many of these ejected materials were deposited locally. "Very proximal" material is found up to 500 km from the crater rim. "Proximal" material is from 500 to 1000 km from the rim, which includes the coast of Texas. Very proximal and proximal material is found in layers up to 80 m thick in Mexico and several meters thick across the Caribbean. It is typical for the ejecta from craters to largely fall back into the crater or within a distance equal to a few crater radii because most of the proximal material is of relatively large size. Generally, 90% of the ejecta falls within five crater radii from the rim of a crater, or about 1,000 km for Chicxulub. The very proximal material likely contributed to generation of tsunamis in the ancient Gulf of Mexico when it fell into the sea.

"Intermediate" layers are centimeters thick and found from the Caribbean, across the United States, and into Canada at distances from 1,000 to 5,000 km from the rim. The intermediate layers contain many millimeter-sized spherules, denoted as microtektites or type I spherules (Bohor and Glass 1995). The intermediate layers also contain tektites that form from molten target rock ejected from the crater. Tektites are often aerodynamically shaped due to their passage through the atmosphere while they were molten. Figure 3.5 shows an exposure of the intermediate layer at Trinidad State Park in Colorado.

One of the notable components of the intermediate layer, as well as more distant layers, is grains of quartz that show colored features under the microscope related to fracture planes in the small grain of rock (Figure 3.6). Such "shocked quartz" grains are one of the principal lines of evidence that the intermediate layers were formed by an impact and not by a volcanic eruption. Since the shocked quartz did not melt, it is a remnant of the pulverized rock. The basal rock at the Chicxulub site is composed of granite, which is the likely source of the quartz. The distribution of the shocked quartz is not uniform over the Earth. It is thought that the pulverized ejecta traveled around the Earth in a westward direction. Morgan et al. (2006) showed that the maximum size of the shocked quartz particles declines roughly as the square root of the distance from the impact site. In the United States and Canada, up to about 2,000 km distant from the crater, quartz particles are found as large as 200 μm in diameter, the size of fine sand. In Europe and New Zealand, about 10,000 km distant, particles are smaller than

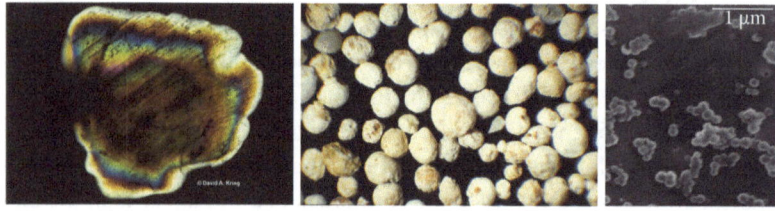

Figure 3.6 Left: a shocked quartz grain as seen under a polarizing microscope, courtesy of David Kring. Center: microkrystites from Caravaca, Spain, courtesy of Jan Smit. Right: an electron microscope picture of 66-million-year-old soot particles from the Distal layer, courtesy of Wendy Wolbach.

75 μm. Particles of these sizes have fall velocities that vary in proportion to the square of the particle size. Therefore, it would be expected that the particle size would decline as the square root of the distance.

The Distal (remote) layer is found across the globe beyond 5000 km from the crater and is about 3 mm thick everywhere. The Distal layer occurs at Chicxulub, and on top of the intermediate layer, indicating that it spread worldwide and remained in the air longer than the other layers. Given its thickness, the mass of the Distal layer is about 3.75×10^{15} kg, which is about twice the mass of a 12 km diameter asteroid. The Distal layer is the key to understanding global effects.

The Distal layer is about the same thickness 7000 km from Chicxulub and 15,000 km away, indicating that particles in it were small enough, or carried fast enough, to be transported globally before they fell out (Smit 1999). The Distal layer is dominated by spherules of rock called microkrystites or type 2 spherules (Figure 3.6), which likely formed from vaporized rock. They are not aerodynamically shaped, suggesting that they did not travel through the atmosphere in a molten state as did tektites. The microkrystites have variable composition between sites because they have been chemically altered by local processes through time. However, they also have varying composition at single sites, possibly indicating that they contain varying mixtures of target material and asteroid. It is also possible that the variable composition is due to varying temperatures of formation and oxygen abundance in the cloud of vaporized ejecta. The microkrystites contain iridium, indicating they contain parts of the asteroid. It is estimated that the mean size of the particles is 250 μm, and there are an astounding 20,000 particles per square centimeter on the ground where the Distal layer is exposed, yielding a mass of 2×10^{15} kg, or half the mass of the Distal layer (Smit 1999).

The origin of the Distal layer is a subject of debate. One theory suggests that the spherules come from material that was vaporized. In this theory, rock vapor expanded thousands of miles (kilometers) above the top of the atmosphere into space, where it cooled to form spheres, which then fell and re-entered the atmosphere at high velocity. When it re-entered, the material heated the upper atmospheric air to temperatures of 1000 or 2000 degrees kelvin. Such hot air and rock would radiate enough infrared energy to the ground to cause global wildfires. In this theory, a lower limit can be set for the energy of the asteroid or comet that will start fires over the entire planet. The lower limit is between 10^8 and 10^9 Mt, very close to the energy of the Chicxulub object (Toon et al. 1997).

The second theory is that the Distal layer spheres did not form from rock vapor but instead formed from the melted target and impactor as the melt was torn apart by friction with the air to form little spheres in the expanding ejecta plume. Then, the spheres were transported across the Earth at altitudes below the start of space as part of the pulverized and melted ejecta. However, this theory is inconsistent with the data on particle sizes. The shocked quartz particles only have 200 μm sizes over a small fraction of the Earth, and the abundance of shocked quartz particles declines very rapidly with distance from the crater. The Distal layer, however, has uniform thickness over Earth, and 250 μm sized particles are found worldwide. It is difficult to understand how one component of the layer could fall out rapidly as the layer spreads over the Earth while a similar sized component does not fall out during the initial spreading.

Because it is a gas, rock vapor will behave very differently than pulverized rock. The vaporized rock will expand due to the pressure gradient across the gas and will rise and shoot into space. Fortunately, we have good observations of the behavior of vapor plumes from multiple impacts from the collision of fragments of the Shoemaker-Levy 9 (SL-9) comet with Jupiter. Carolyn Shoemaker, Gene Shoemaker, and David Levy discovered the comet after it broke into pieces due to Jupiter's gravity. Fragments of SL-9 in the 1 km diameter range hit Jupiter in 1994, making many observed plumes (Boslough and Crawford 1997). There are several differences with the Chicxulub case since comets like Shoemaker-Levy 9 produce plumes mainly of water vapor, rather than rock vapor, and Jupiter does not have a solid surface for an impact like Earth does. The Chicxulub impact energy was also at least 100 times greater than that of the SL-9 fragments. Nevertheless, similar behavior of the vaporized material is likely.

Because the Jovian vapor plumes from the impacts were large compared with the characteristic depth of the pressure decline with altitude in Jupiter's atmosphere, the bottom of the vapor plume was at much higher pressure than the top. The pressure gradient in the hot plume caused it to accelerate upward along the impactor entry trajectory where the air had been pushed aside. The plume expanded into space, cooled, and reached several thousand kilometers about the Jovian clouds within about nine minutes. The remnant plume cooled, then collapsed downward, leading to compressional heating of the lower atmosphere to temperatures of at least 2000 K after about six minutes. Following this collapse, the plumes rose and fell several times, as they floated thousands of kilometers across Jupiter's atmosphere. Several impacts had rings of hot gas that expanded over large regions. At about two hours after the impact, one ring had expanded over an area about the same as the Earth's area and had a temperature near 600 K.

A major difference between the Chicxulub impact and that of SL-9 is that the Chicxulub plume contained rock vapor. Johnson and Melosh (2012) modeled an expanding rock vapor plume from an asteroid impact on Earth at a velocity of 21 km/s. As the vapor expanded and cooled, it reached temperatures and pressures at which molten rock particles nucleate from the gas. Following the formation of these particles, rock vapor condensed on them, and they grew to larger sizes. These processes are all like ones that occur when water vapor condenses to form a cloud. Eventually, the rock vapor was depleted due to growth of rocks and the expansion of the vapor cloud, and the rock particles ceased growing because there were few molecules of rock vapor left to collide with. About 1000 s after the impact, the vapor cloud was about 19,000 km in diameter, much larger than Earth, whose diameter is about 12,000 km, and the fastest moving portion of the cloud was still expanding at about 19 km/s at an altitude more than 9,000 km above Earth's surface. Parts of the plume had temperatures of 1600 K at this time, and the plume had a low density so that particle growth stopped. Even after 10,000 s (almost 3 hours), about 44% of the rock remained as vapor in the model. The average-sized rock particle in the model had a diameter of 232 μm, about the same as the observed size.

As mentioned previously the spherules compose about half of the Distal layer. The spherules are obvious because of their shape and large size. But what composes the rest of the layer? Unfortunately, it is difficult to answer this question because the material in the layer has been chemically modified in the last 66 million years. However, we know of two components. First, and most surprising, are carbon particles, soot, some of which are pictured in Figure 3.6. These particles are composed of small carbon spheres that are linked together to form chains of particles. Similar particles are seen in smoke from modern forest fires. The surprising thing about these particles is that they are so abundant that it would require burning essentially all of the exposed biomass on the Cretaceous Earth to produce them. A second type of particle found in the Distal layer is very small iron particles. These are relatively easy to detect among all the materials in the Distal layer. They are similar in size to the small carbon spheres seen in Figure 3.6, several hundredths of a micrometer. These iron particles are widespread. Unfortunately, their abundance has not been quantified, so we do not know the fraction of the Distal layer they compose. Possibly these small iron particles are remnants of the rock vapor cloud that did not condense to form the larger microkrystites. There are other likely components of the Distal layer that have never been clearly identified and quantified. The presence of shocked quartz suggests that there are other small particles composed of crushed rock. Estimates of the possible mass of these particles cover a wide range.

3.4 Gases Released from the Impact

When rocks are shocked to high pressures and temperatures, they can release gases such as carbon dioxide (CO_2), sulfur trioxide (SO_3), and sulfur dioxide (SO_2). CO_2 is a potent greenhouse gas. SO_2 is a common industrial and fossil fuel product and a contributor to acid rain. After a very explosive volcanic eruption, SO_2 from the volcanic magma can enter the stratosphere, where it reacts with gases already in the atmosphere to produce sulfuric acid particles. These stratospheric volcanic sulfate particles have historically caused significant climate changes on Earth, and some scientists are proposing that we purposefully inject SO_2 into the stratosphere to produce sulfuric acid particles that would cool Earth by reflecting sunlight back to space and offset the global warming due to human releases of greenhouse gases. Such "geoengineering" would come with many risks and concerns (e.g., Robock 2020).

Based on drill cores in areas outside the Chicxulub crater, the sedimentary rock at Chicxulub contains 30–50% sulfur-rich evaporites, as well as carbonates. Carbonates are rocks that formed ancient reefs built from the shells of plankton that drifted downward to the sea floor when the plankton died. Limestone is an example. Evaporites are rocks that form when water bodies evaporate. Sediments near Chicxulub include evaporites such as anhydrite, which is anhydrous calcium sulfate, as well as gypsum, which is calcium sulfate dihydrate. Gypsum is well known as a component of wall board often used in the interiors of houses. However, cores of the Chicxulub crater show that less than 1% of evaporites remain after the impact (Gulik et al. 2019). Artemieva and Morgan (2020) used calculations of the pressures applied to the sediments during the impact to estimate that 325 ± 60 Gt of S and 425 ± 60 Gt of CO_2 were released by the high-velocity impact from the anhydrite and carbonate. This mass of CO_2 is only about 20% of the CO_2 currently in Earth's atmosphere, and atmospheric CO_2 was more abundant at the end of the Cretaceous than now. So, this injection would have produced only a small additional climate warming. The mass of sulfur, however, is immense compared with that from volcanic eruptions. For example, the sulfur emitted is about 15,000 times that from the Mt. Pinatubo volcanic eruption of 1991, the largest of the twentieth century, which produced an easily measured global cooling. The mass of sulfur estimated to come from the target rock is about 7% of the mass of a 15 km diameter asteroid. It is thought that the asteroid was a carbonaceous chondrite, type CM2 (Figure 3.2). Such asteroids contain about 3% sulfur. Over the range on chondrites, which are 85% of meteorite falls, sulfur varies from 1.6 to 5.7%. Therefore, the amount of sulfur coming from the asteroid itself would likely be similar to that evolved from the sediments. The high sulfur content of asteroids makes the claim that the impact site was special because of the sulfur-rich sediments unlikely to be a significant factor in explaining why other large impacts have not so far been found to be linked to extinctions. Several other gases might have been injected from the target or the asteroid in significant quantities, including water, chlorine, bromine, and iodine, which might be important for destroying the ozone layer. Large quantities of seawater may also have been splashed into the atmosphere by the impact.

It is difficult to determine the amount and form of sulfur in the Distal layer because there is so much sulfur in seawater and in the Earth environment. An important issue is whether the sulfur released in the impact remained in the atmosphere after the first few hours or was converted back into minerals while the ejecta and vaporized rock were traveling through the atmosphere. It is likely that some sulfur gases released from the asteroid and target traveled with the vaporized rock. As the rock vapor condensed to form the microkrystites, it is likely that the sulfur reacted on the rock surfaces and was incorporated inside the particles. If all the sulfur were removed this way, separate particles of sulfate would not form, and sulfur from the asteroid would play little role in the aftermath of the impact. The sulfur vaporized from the pulverized and melted target rock would likely travel with the pulverized and melted ejecta. It has recently been found from laboratory studies and observations of volcanic clouds that sulfur dioxide will react on the surfaces of volcanic ash, pulverized rock from the eruption, very quickly. The reactions are limited by the exposed area of the rock. It is expected that the pulverized and melted ejecta have 100 times or more mass than the asteroid, so the surface area of the rock may have been large. Therefore, the sulfur may also have been lost to the ejecta rather than forming independent particles. Simulations of the process of sulfur dioxide interacting with the ejecta have not been carried out to date, so we don't have a quantitative idea of how important independent sulfates might have been.

In summary, there are several assured destruction phenomena due to asteroid or comet impacts, such as tsunamis, earthquakes, and shock waves, but each of these is likely to have extended over only a small fraction of the Earth. The possible global destruction mechanisms all relate to the composition of the Distal layer. We know that half of this layer is composed of microkrystites, 250 μm sized rock spheres. We also know that the layer contains a large amount of soot. In one set of hypotheses, the microkrystites started fires globally when they re-entered the atmosphere, and the soot is proof that these fires occurred. To set global fires, a minimum impact energy close to that from the Chicxulub object is needed. Therefore, global fires are not expected from objects that are much smaller. As we shall discuss, the fires and the soot are able to explain both the extinctions on land and those in the oceans. An alternative theory is that sulfate particles coming from the sulfur degassed from the evaporites that formed the target of the impact blocked sunlight, creating such cold temperatures that extinctions occurred on land. However, this theory lacks an explanation of the extinctions in the oceans. Of course, many devastating phenomena occurred after the impact, and each may have played a role in the events on that day and in the subsequent extinctions. A single explanation for everything isn't necessary.

Chapter 4
Worldwide Fires Killed the Dinosaurs

We have a murder mystery! What killed the dinosaurs as well as about 75% of all the species on the planet? Like many murder mysteries, we have clues, some of which might be false leads, but we have no eyewitnesses. Many suspects have been accused since the murder was first discovered. Some of them have alibis and have been eliminated.

4.1 Clues

Several groups used the paleontological record to look for clues about the Cretaceous-Paleogene (K-Pg) extinction. Dale Russell (1979) summarized what was known in the years just preceding the discovery of the asteroid impact. He suggested that there "was no evidence of a decline in diversity or of a meridional restriction in the geographic distribution of large reptiles toward the end of Cretaceous time." He also found that "it is probable that the time scale of biotic and geomagnetic events during the . . . K-Pg . . . transition is at the limit of stratigraphic resolution." In other words, the extinction event was sudden, and the dinosaurs were doing fine right up until the event. Considerable debate in the paleontology community occurred in the past few decades, and continues today, about whether dinosaurs were fading away before the K-Pg boundary or not. Dinosaur fossils are not common, making it difficult to study the abundance and diversity of non-avian dinosaurs through time. However, dinosaurs are just the most famous of the creatures that became extinct. Seventy-five percent of marine species went extinct at the K-Pg boundary. Some of these, such as plankton, left abundant fossils. There is no evidence that life in general was suffering unusual stress before the impact.

Russell went on to hunt for clues about the cause of the sudden extinction by examining the number of genera of various organisms before and after the extinction. In taxonomy (Figure 4.1), the most basic level is the species. For example, our species is *Homo sapiens*. However, species do not last long in the geologic record, so a species extinction event is difficult to recognize against the constant turnover of species. The next taxonomic level is the genus. We are the only living members of our genus, *Homo*. However, in the recent past there were others in our genus such as Neanderthals and Denisovans. The next

Earth in Flames. Owen Brian Toon and Alan Robock, Oxford University Press. © Oxford University Press (2025).
DOI: 10.1093/9780197799734.003.0004

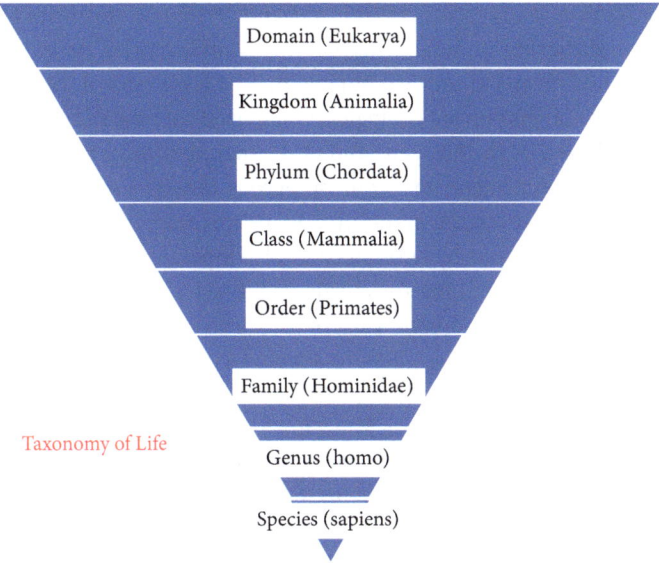

Figure 4.1 The organization of life, with humans as an example.

level of taxonomy is the family. We are in the family of great apes, *Hominidae*, which includes four genera: *Ponga* (for example, orangutans), *Gorilla*, *Pan* (for example, chimpanzees), and *Homo*. Russell compared the number of genera in various groupings of families before and after the K-Pg extinction event. He concluded that freshwater environments were relatively little affected, while marine environments were strongly impacted. He traced the large effects in the oceans to significant declines in phytoplankton and those whose food source depended on phytoplankton, while bottom-dwelling detritus consumers were less impacted. Detritus is dead particulate organic matter, so bottom dwellers may have been well fed just after the extinction event. In the oceans, different types of phytoplankton responded differently; coccoliths and foraminifers were strongly impacted, whereas diatoms and radiolarians were not. Some have interpreted these differences as being related to the composition of the shells of these plankton. Coccoliths and foraminifers make their shells mainly from calcium carbonate (limestone), while diatoms and radiolarians mainly make their shells from silica. It has been suggested that the ocean may have become acidic, which would quickly dissolve the calcium carbonate but not the silicate. However, there are many other differences between these organisms such as their ability to survive in dormant forms.

After the discovery of the asteroid impact, Doug Robertson et al. (2004) used commonalities in the behaviors of the survivors to suggest mechanisms for the extinction. They examined non-marine vertebrate survivors and concluded "that sheltering underground, within natural cavities, or in water was the fundamental means to survival during the first few hours of the Cenozoic." (The Cenozoic, Earth's current geological era, started with the Paleogene period, which was followed by the Neogene and Quaternary periods.) Sheltering might be needed to protect organisms from numerous threats such as exposure to ultraviolet light. However, the major immediate threats were likely global fires, discussed below, followed by prolonged cold temperatures and lack of food. Robertson et al. suggested that the survivors all had small enough body sizes to find shelter from cold and fires. Most of the survivors may have had lifestyles that favored sheltering, so they were pre-adapted to survive. For example, many mammals may have lived in holes in the ground. It is common now for mice and other small creatures to avoid a forest fire by staying underground while the fire passes. Likewise, many creatures now survive cold weather by sheltering. Snakes and rodents pass the winter in underground dens. Even 1 m (a few feet) of soil is enough to provide sheltering. Alligators have hollows in the banks of rivers where they wait for winter to end. Frogs and turtles bury themselves in mud to escape cold weather. Many insects live entirely underground or have life stages that are underground. Also, many plants have seeds underground or roots that can propagate. Organisms that burrow may also have access to buried plant foods, such as roots or insects that live underground. Not all the dinosaurs died in the K-Pg event. Some avian dinosaurs, the ancestors of modern birds, survived. It is possible that they survived because they lived in holes in trees, or cliffs, as do some modern birds. Some of the avian dinosaurs also lived near seashores, or swamps, which may also have provided protection and food.

Robertson et al. (2013) examined survivors in the oceans and freshwaters to consider why extinctions were more widespread in the oceans than in freshwater environments. Neither water environment would be influenced by fires. However, the marine data suggest the organisms in the photic zone were less likely to have survived than those whose lifestyles did not depend on sunlight. The photic zone in the oceans is a shallow region around 100 m deep in which light levels exceed 1% of those at the surface. Such light levels allow for photosynthesis. Robertson et al. argued that freshwater organisms are not highly dependent on photosynthesis because they are supplied with food from detritus washing into lakes and streams from the land. There is also a store of detritus at the bottom of the oceans and a bounty of organisms that normally lives by eating this material, which may explain why deep-water organisms were less impacted than those living in the photic zone of the oceans. Thermal stresses in the ocean may have also played a role. Even now small changes in surface temperatures are challenging for corals. Many creatures move significant distances to escape small temperature changes due to phenomena such as El Niño. However, with global-scale cooling at the surface, there would be no place to move. Freshwater organisms are exposed to temperature fluctuations yearly and have adapted several survival mechanisms such as burial in mud. It is possible that warm groundwaters flowing into rivers may have also provided some locations in freshwaters that were not much colder than normal.

4.2 Theories for the K-Pg Extinction

Many ideas have been proposed for the K-Pg mass extinction. Brian's favorite was proposed by the cartoonist Gary Larson. His cartoon pictures a group of tough-looking dinosaurs puffing on cigarettes with the title "The real reason dinosaurs became extinct." Brian likes this theory because it is testable. There are no cigarette butts found at the K-Pg boundary, so this extinction theory is wrong. Even though the idea is not plausible, it is certainly amusing.

Several extinction mechanisms focus directly on dinosaurs and fail because they do not explain extinctions of plants or extinctions of plankton and other creatures in the oceans. As an example, it was hypothesized that mammals outcompeted dinosaurs or ate their eggs, driving them to extinction. Dinosaurs arose following the greatest known extinction event 250 million years ago at the transition between the Permian geologic Period and the Triassic Period. Mammals also arose during the Triassic, and during much of this period they were probably nocturnal, which helped to avoid predators. Dinosaurs became the dominant large land creatures about 200 million years ago after another mass extinction event at the end of the Triassic period and the start of the Jurassic Period. While dinosaurs were the dominant large animals on Earth, mammals dominated the habitats available to small creatures. So, dinosaurs and mammals coexisted for more than 200 million years, and there is no evidence that mammals suddenly started to compete with dinosaurs for the large creature niche. Even if they had competed effectively, it would not have caused extinction in the oceans. Another example in this genre of dino-centric extinction theories is that dinosaurs died from an infectious disease. But a pandemic affecting dinosaurs is not likely to cause extinctions of plants or plankton, so it does not explain the wide scale of the K-Pg mass extinction.

Another class of extinction mechanism depends on global-scale changes in environmental conditions and climate. However, climate and sea level are usually slowly varying and might be avoided by moving or adaptation. It is difficult to imagine global-scale extinctions due to a slowly changing environment unless the entire environment became uninhabitable. Also, the evidence suggests the K-Pg extinction was sudden.

Mass extinctions in the ocean could occur if the ocean's oxygen content fell. Even today there are many areas where low oxygen in the water leads to fish kills. For more than half of Earth's history, up until the great oxidation event about 2.2 billion years ago, atmospheric and oceanic oxygen were very low. Oxygen now composes 20% of the atmosphere. Prior to 2.2 billion years ago, oxygen may have only contributed a few molecules out of a million molecules of air, the remainder being mostly nitrogen. Life on Earth then consisted of simple creatures that did not need oxygen to survive, such as bacteria, sometimes called blue-green algae despite not being algae. Surprisingly, photosynthesis, which produces oxygen, was developed long before oxygen levels rose. The delay in having free oxygen in the atmosphere and ocean was partly because it took a long time to oxidize all of the iron and sulfur in the environment. However, photosynthesis does not actually produce net oxygen because oxygen is consumed when organisms decay after death. Instead, oxygen is produced when organic carbon is buried so that oxygen cannot be consumed during decay. Another way to produce oxygen occurs when hydrogen from water escapes the planet to space, leaving oxygen behind. These are very slow processes.

Currently, oxygen is supplied to the deep ocean mainly when cold oxygen-rich water at northern high latitudes sinks to the ocean floor and spreads globally. Without that, the deep ocean would be deprived of oxygen. Very cold poles and polar ice sheets like we have today occupy only a modest fraction of Earth's history. Probably ice sheets have been present for about 20% of the past 600 million years when complex life was present. There were no ice sheets when the dinosaurs became extinct, and indeed some dinosaurs lived on Antarctica. Therefore, the supply of deep water was more restricted in much of Earth's history than now because it was not so cold at the poles.

Mass extinctions only occurred about 4 times in the past 600 million years. A leading example of this type of extinction event is at the boundary between the Permian and Triassic periods, 250 million years ago. Evidence points to extensive volcanism increasing carbon dioxide levels, which was followed by very warm temperatures, little deep-water formation, and an anoxic (with no dissolved oxygen) ocean causing extinctions in the oceans. Many related theories have been developed. However, at the time of the K-Pg extinction, there is no evidence of an anoxic ocean, of significant long-term climate changes, or of changes in sea level.

One major contender for the K-Pg extinction has been volcanism. Many volcanoes are explosive due to their viscous lava and are familiar to most people due to the destruction they cause locally and the climate changes they cause by injections of sulfur gases into the stratosphere. Such volcanoes are found along the "ring of fire," a zone where ocean crust is being subducted under continental plates. The Earth's crust is separated into many plates, and the subduction regions mark areas where oceanic plates are being destroyed by being shoved under continental plates, and continental plates are being built by volcanism. The ring of fire currently encircles the Pacific Ocean with huge numbers of explosive volcanoes, including famous ones such as Mt. St. Helens in the US state of Washington, Tambora on Sumbawa, Krakatau in the sea between Java and Sumatra, El Chichón in Mexico, Mt. Pinatubo in the Philippines, and the many other volcanoes of the Pacific Northwest, Alaska, the Kamchatka Peninsula, Japan, Indonesia, New Zealand, and South and Central America.

In contrast to explosive eruptions, basaltic eruptions form lava that is not very viscous and can flood out over great distances, sometime from long fractures in the crust rather than from individual mountains. Such flood basalts are found along the mid-ocean ridges where new ocean crust is forming and is spreading apart along plate boundaries such as in Iceland, where the mid-ocean ridge is exposed at the ocean surface. Flood basalts also occur in locations where there is a plume of lava rising from the mantle, such as in Hawaii. Flood basalts are an extreme type of eruption that creates vast overlapping sheets of volcanic lava. One such example is the Siberian Traps whose formation may have caused the Permian-Triassic extinction. Some scientists have argued that the Deccan Traps, located in current west-central India, caused the K-Pg extinction. The word "trap" comes from "*trappa*," a Swedish word for steps or staircase. Flood basalts repeat over long periods of time, building up layers that can look like gigantic steps. The Deccan Traps are one of the largest volcanic features on Earth.

At the time of the K-Pg extinction, India was a subcontinent about halfway between Madagascar and its current location. As India moved, it passed over a mantle hotspot, much like the Hawaiian Islands have done. Extensive layers of lava built up over a period of millions of years. Currently the volume of lava fields is about 1 million cubic kilometers, and before its erosion it would have been even bigger. Even its current mass is about 1000 times the mass of the Chicxulub asteroid or the Distal layer. Its mass is enough the cover Earth in a layer of lava about 1 m thick.

Initially volcanologists claimed that the iridium layer was due to the Deccan Traps, but volcanoes do not usually vent iridium, and the iridium containing material thickens toward Mexico, not India. Adherents to the Deccan Trap theory of the K-Pg extinction have now moved to the extinctions having augmented the effects of the impact or prolonged them. It has been suggested that the amount of sulfur emitted might have caused climate changes, but flood basalts do not usually inject sulfur into the stratosphere where it might linger long enough to alter the climate. Also, the Deccan Traps formed over millions of years, and this period does not show a prolonged cooling. Currently there is no evidence for a short, but extreme lava flow that correlates with the K-Pg boundary. Also, it has been argued that the sulfur might have acidified the oceans, causing extinctions of plankton with shells made of calcium carbonate. However, it is difficult to imagine enough sulfur being emitted to acidify the world ocean. Moreover, there are other flood basalts, comparable to the Deccan Traps, that did not cause extinctions. The Columbia River Basalt in the United States which mainly erupted from about 15.6 to 16.7 million years ago and has about 10% the volume of the Deccan Traps is not associated with an extinction. Some scientists theorize that the Deccan Traps may have made the recovery from the K-Pg extinction longer or more difficult. Others suggest the Deccan Trap eruptions may have set the stage for the extinction from the impact (Green et al. 2022). However, there is little evidence that the Deccan Traps caused the extinction. Nevertheless, supporters of this theory continue to pursue it. That is how science is supposed to work. If a new idea arises, such as an asteroid killing the dinosaurs, everyone tries to challenge it, find new data that either supports or contradicts it, and then, if necessary, come up with alternative ideas that explain all the data.

4.3 The Impact Theory

Several scientists, including Harold Urey, who won the Nobel Prize for Chemistry in 1934 for discovering deuterium, proposed before Alvarez et al. (1980) that asteroid or comet impacts were responsible for mass extinctions (Urey 1973). But their proposals were ignored because they did not produce any direct evidence of a link between impacts and extinctions. In contrast, the Alvarez group produced data showing an impact and a mass extinction occurred within a year or so of each other—an infinitesimal period in the geologic record. It is now generally accepted that the asteroid caused the extinction (Schulte et al. 2010), though the details of the mechanism for the extinction are still hotly debated.

As we discussed in Chapter 3, several effects occurred within a few thousand kilometers of the impact crater, such as tsunamis, earthquakes, and atmospheric shock waves, but none of these is likely to have produced a global extinction. Rather, something in the Distal layer is the likely cause of the extinction. The Distal layer contains three obvious items, as illustrated in Figure 3.6, and some items that are not so obvious, which likely caused the extinction.

The sulfate component of the Distal layer is not obvious because sulfur is so common in the environment. Many scientists think the sulfur release was a critical part of the extinction, but there are two major unresolved issues. First, the sulfur-containing particles probably were composed of sulfuric acid solutions. Sulfuric acid solutions are transparent liquids. They look like water. Therefore, although the sulfuric acid particles, like an everyday water cloud, could reflect sunlight back to space and prevent it from reaching the surface, they would not absorb much sunlight. Absorption of sunlight is a much more powerful way to reduce the amount of light reaching the Earth's surface than reflection of light. Calculations by Tabor et al. (2020) (Figure 4.2) of the climate change following an injection of 325 Gt of sulfur as SO_2 suggest that the amount of sunlight reaching the surface after passing through the veil of sulfuric acid coming from the impact would still be about 20% of normal, about the same as now occurs on a very cloudy day. Such losses of sunlight, which are extreme from the point of view of the climate, are unlikely to cause a mass extinction in the ocean because plenty of light remains for photosynthesis. Most people who favor the sulfate idea invoke dust or soot in the sulfate to add absorption and further reduce sunlight. The second problem is that sulfur gases are very reactive. As the rock vapor plume cooled and solid or liquid rock reformed, it is likely that much of the sulfur would react back onto the rocks from which it came, forming minerals such as calcium sulfate (gypsum) or magnesium sulfate (Epsom salts). If so, much of the sulfate would be removed within a few days on the large spherules found in the K-Pg layer, shown in Figure 3.6, and have little impact on the climate. However, no one has investigated the details of these interactions to date.

The shocked quartz grains, shown in Figure 3.6, represent portions of the target materials that were pulverized. It is likely that this rock dust played a role in blocking sunlight from reaching the surface. Rock can absorb light, making it efficient at blocking sunlight, but rocks are dense, and many were so large they may have fallen out of the atmosphere in days to months.

Unfortunately, the rock dust from the pulverized ejecta is poorly quantified, and the shocked quartz is just a minor component. There is a component of the Distal layer composed of very small iron particles. Although the size of these particles is known, their mass has not been quantified. They might represent parts of the vaporized impactor and target that did not get swept up as the spherules in the layer formed as described in Chapter 3 and therefore remained as small dust particles. Calculations suggest about 44% of the vaporized material was left as rock vapor when growth of the spherules stopped. Future studies of this material would be valuable. Calculations of the climate impact of injecting 2000 Gt, about half the mass of the Distal layer, of very small iron rich particles show that the dust can initially reduce sunlight to essentially zero at the surface (Figure 4.2). However, the dust particles rapidly coagulate to large sizes, and 99% falls out within 3 months. Light levels below 1%, which are too low for photosynthesis, persist for about 45 days. Such prolonged darkness could be a blow to phytoplankton and the marine food chain. Under normal conditions, the marine food chain recycles all of its carbon in about a week, unlike the food chain on land, which cycles carbon over about ten years. Therefore, 45 days in which phytoplankton could not generate new organic carbon could be an issue

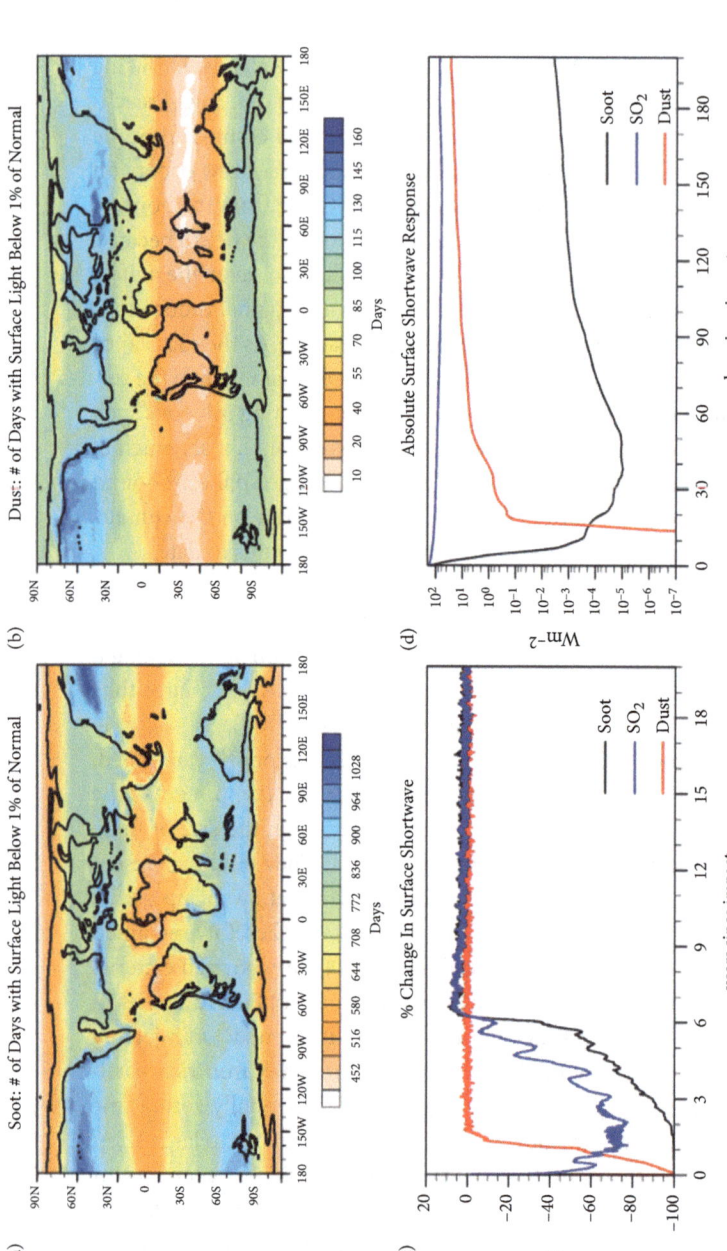

Figure 4.2 Light levels following injections of sulfur dioxide, soot, and dust by the impact at the K-Pg boundary. Figure 4.2c, d shows that only dust and soot reduce light levels below 1%, the boundary where photosynthesis ceases. In the case of dust, the injected mass is so large that the particles quickly coagulate to produce large particles, which fall out of the atmosphere relatively quickly. As a result, low light levels only occur for about 6 months. However, soot particles remain relatively small due to their smaller injected mass. As a result of the small size, and low density of the soot particles, they remain airborne long enough to reduce light levels below the limit of photosynthesis for about 2 years on a global average and 1 year even at locations in the tropics. Figures from Tabor et al. (2020).

in a wide range of the oceans, but plankton survive long periods of darkness during winter near the poles. As shown in Figure 4.2, tropical light levels too low for photosynthesis only persisted for a week or two due to a dust injection.

The spherules illustrated in Figure 3.6, which constitute about half the mass of the Distal layer, are about the same size as beach sand. As any beachgoer is likely to know, sand falls to the ground in seconds when thrown into the air. The spherules had to fall about 60 km to reach the ground, so it probably took a few days for them to be removed from the atmosphere. If any dinosaurs were still alive when the rain of spherules reached the ground, it probably felt like being pelted by sand, which would not injure the dinosaurs. Given their short residence time, the spherules would not have a direct impact on the climate. However, they are likely to have killed most, if not all, the large dinosaurs by starting fires across the Earth, which broiled them.

The evidence for global fires is also present in the Distal layer. Anyone with a wood-burning fireplace will know there are two types of materials left from a fire that contain a lot of black carbon. One component is charcoal, which is mostly present in the ash or remnants of the wood. Charcoal is made of large particles, which will not travel far from the fire. The second component is made of very small particles, tens of nanometers in size, which clump together to form chains and sheets of particles. In the fireplace, these particles coat the chimney with black soot. Both charcoal and soot are found in the K-Pg Distal layer (Wolbach et al. 1988). Figure 3.6 illustrates some soot remaining after 66 million years in the Distal layer. The soot mass is roughly the same in the 11 places where it has been measured, indicating that it spread globally (Wolbach et al. 1990). The presence of charcoal in these same areas indicates that the fires were also widespread. Some have argued that the smoke came from the crater that is now in a hydrocarbon-rich area, but that is inconsistent with the presence of charcoal on the other side of the planet. Only a small fraction of the soot could have come from the crater. We know the Distal layer took only a few years to form because that is the longest time it would take the smallest particles in the layer to fall out of the atmosphere. Some have argued that the soot came from fires that burned long after the extinction event when dead and dying forests went up in flames. In that case, one would expect the soot to lie on top of the Distal layer. Figure 4.3 shows the soot is within the Distal layer, with the highest abundance at the base of the layer just like iridium, so the soot was created during the period when the Distal layer formed, not later. The mass of soot and charcoal in the layer is about 56 Gt, which is about 2% of the mass of the Distal layer. About 26% of the material, roughly 15 Gt, is small soot particles rather than charcoal.

That is a lot of soot. It is estimated that the above-ground biomass on end-Cretaceous Earth was like modern tropical forests, about 2 g per square centimeter, which is about 3×10^{18} g of fuel over all the continents. Measurements of soot production from burning modern forests show that about 0.5 g of soot is

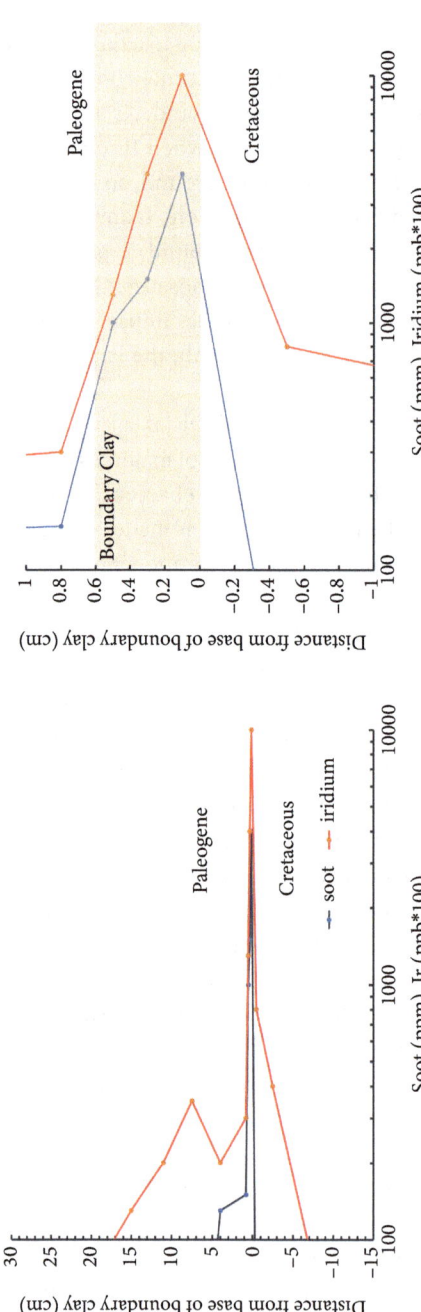

Figure 4.3 The distributions of soot and iridium in the boundary layer, made mostly of clay, marking the K-Pg extinction at Woodside Creek, New Zealand. The figure on the left has similar distributions of iridium and soot. Each is enhanced in the Boundary clay by factors of 100 or more relative to the sediment deposited in the Cretaceous. On the horizontal axis, ppb*100 refers to the ratio of the mass of Ir multiplied by 100 to the mass of clay measured in parts per billion, while ppm refers to the ratio of soot to clay measured in parts per million. The figure on the right shows iridium and soot are closely related within the Boundary clay, which is about 0.6 cm thick. Drawn using data from Wolbach et al. (1988).

produced from burning 1 kg of fuel. Laboratory studies show that the emissions of soot rise by about a factor of 10 when fires are oxygen-starved as compared with fires in which oxygen is not limited. Assuming the global biomass was on fire, transport of oxygen horizontally into the fire would not occur in most places, and it is likely that the K-Pg fires were oxygen-starved. The soot mass is roughly the same in the 11 places where it has been measured. If we divide the amount of soot by the continental area of Earth (1.5×10^{16}g/1.5×10^{18} cm^2), we find that about 0.01 g of soot must have been generated from each square centimeter of continent on average. Multiplying the soot emissions for oxygen-starved fires times the global biomass provides 15 Gt of soot, just the amount needed to account for all the soot observed.

In summary, the amount of soot found in the Distal layer is consistent with oxygen-starved fires burning essentially all the material above ground. Probably some below-ground biomass also burned. Most of Earth's land surface must have been on fire in the hours and days after the K-Pg impact.

4.4 Whole World on Fire

How did all these fires start? This was a major puzzle in the years just after the Alvarez et al. (1980) discovery of the impact and Wendy Wolbach and coworkers' (1988) discovery of soot in the Distal layer. The puzzle was solved in a series of papers by Jay Melosh et al. (1990). They recognized that when the spherules fell back to Earth from the vaporized rock plume that extended thousands of kilometers above the surface, they would carry a vast amount of energy, as discussed in Chapter 3. Initially the energy would be kinetic due to the velocity of the spheres. Melosh et al. assumed the spherules would have about one-ninth of the energy of the asteroid. As the spheres re-entered the atmosphere, they would be heated and slowed by friction with the air.

This same process currently happens as shooting stars enter the atmosphere. Probably everyone has seen shooting stars as bright streaks of light zooming across the nighttime sky. Depending on ambient light levels, cloudiness, time of day, and other factors, one might see a shooting star every 10 to 15 minutes on a typical night. During the peak of a meteor shower, a shooting star could be visible once a minute. About 40 tons of shooting stars, which are mostly about the size of a grain of sand, enter the atmosphere every day, but generally we do not find them on the ground. Due to their high speed, shooting stars are heated by friction to such high temperatures that they evaporate and form a "smoke" of nanometer-sized rock particles pervading much of the atmosphere below about 80 km altitude. There are layers of metal ions that are easily observed near 80 km altitude where shooting stars typically flare up and evaporate.

An individual spherule, like a shooting star, would stop due to friction with the air when it has passed through a mass of air similar to its own mass. The spherules are slightly larger than shooting stars, so they would stop at slightly lower altitudes than shooting stars, probably in the vicinity of 70 km above the surface. However, the spherules did not enter with a high enough velocity to reach temperatures that would cause them to vaporize. Instead, they remained as hot particles. The mass of the Distal layer greatly exceeds the mass of the atmosphere at 70 km. So, the Distal layer fell at the terminal velocity of the spherules, the velocity set by a balance between gravity and friction with the air. The terminal velocity of the spherules is near 2.5 m s^{-1}. Probably, the descent was highly turbulent since the particles would collide with each other. At this speed, it would require about two hours for the particles in the Distal layer to fall downward to reach a mass of air like the mass of the Distal layer, and eight hours to reach the ground.

Unfortunately, some studies assumed that the mass of the Distal layer equaled that of the atmosphere above 70 km since the individual particles slowed to terminal velocity around 70 km when they went through an air mass equal to their own mass. Instead, the mass of air above 70 km is less than 10% of the mass of the Distal layer. So about ten spherules went through the same air when they were slowing down. This error led to the idea that the Distal layer transferred its heat to the air and then cooled and fell below the hot air. The spherules would then block some of the radiation from the hot air reaching the surface. In fact, the suspended spherules forming the Distal layer probably radiated energy downward for about two hours, the time it took to fall from 70 km to 50 km, before they transferred much heat to the air.

Calculations of the energy flux to the Earth's surface are complicated by absorption of light in the atmosphere above the ground and uncertainty about the time during which the spherules entered the atmosphere. Simulations (Toon et al. 1997, Goldin and Melosh 2009, Belcher et al. 2015) suggest that a pulse of light occurred lasting about an hour and reaching peak intensities of about 12–36 kW m^{-2} (about 0.3–0.9 calorie per centimeter squared per second). This amount of energy is about ten times the peak sunlight that reaches the ground and about twice the energy density of an oven set on broil. It is lower than the energy in the light flash from a nuclear weapon that starts fires, but the duration of the nuclear light flash is only a few seconds.

These energy densities and durations are likely high enough to produce fatal burns in most animals exposed to the light. The entire sky would be glowing red like the bar in a broiler oven or a sheet of lava. There would not be shadows, so the only escape would be underground, underwater, or in a cave.

The dinosaurs likely burned to death under the glowing skies. You can do an experiment at home to simulate the death of the non-avian dinosaurs. Go to the grocery store and buy a dead dinosaur. A chicken or turkey will do. Take it home, turn on the oven's broiler setting, and toss the dinosaur into the oven. You may have baked a dinosaur in the past without burning by using a lower temperature setting on the oven. However, with the oven set on broil, the dinosaur will burn, especially if it still has feathers. Such burns would likely be fatal, had the chicken not already been dead when you put it in the oven.

This experiment involves a type of dinosaur that lives in the modern era, a bird, or avian dinosaur. Only the non-avian dinosaurs perished at the K-Pg boundary. Unlike the large non-avian dinosaurs that could not easily hide from the glowing skies, some of the avian dinosaurs must have been small enough to hide in a cavity, such as a hole in a tree, or could stay underwater for a long period.

The radiation from the glowing skies would also be sufficient to set forests and grasslands on fire. Experiments in which heat fluxes in the range of 12 to 36 kW per meter squared were used for just about 8 minutes showed that various forest fuels would ignite into flames according to Belcher et al. (2015). Forest fires usually start in thin materials such as leaves, pine needles, and grasses. Once these materials are on fire, the fire spreads to thicker materials such as branches and vines. Many large animals can run out of forests that are on fire. However, that would be unlikely in the case of the K-Pg fires since fires would start simultaneously over large areas. Only smaller creatures who were able to hide locally, such as underground or in a body of water, would be likely to survive the fires.

While glowing skies and forest fires might have killed many creatures, and possibly even caused the extinctions, there were also likely refugia. For example, swamps would not burn easily and could have provided shade from glowing skies. It might be that the spherules did not enter the atmosphere at high speed over the entire globe, so some regions did not burn. Possibly dense cloud cover and rainfall might have provided some shielding. The degree of fire spread is difficult to assess. However, the fires did not kill creatures just by burning them but also indirectly by injecting vast amounts of smoke into the atmosphere.

4.5 Darkness at Noon

Broiler oven skies and global wildfires do not explain extinctions in the oceans. Another mechanism is needed. The soot found at the K-Pg boundary layer provides a mechanism for extinctions in the oceans. Figure 4.2 shows the light levels expected if all the soot in the K-Pg layer (15,000 million metric tons) is put into the air over the planet. Light levels in Figure 4.2 fall below 1% for these levels of soot and remain below 1% for about 2 years on average. Photosynthesis normally does not occur with light levels below 1% of normal. In the current oceans, the photic zone extends to a depth of about 100 m, and its bottom is marked by light levels falling below 1%. The photic zone is rich with life that is living off the organic carbon produced from carbon dioxide by photosynthetic plankton.

The ocean food chain, unlike that on land, has an equal mass of consumers and producers. The time it takes to produce and to eat the edible organic carbon in the oceans is about ten days. When food production is stopped due to photosynthesis ending, the zooplankton will quickly eat all the phytoplankton, the fish will eat the zooplankton, and the food chain will collapse. This sequence of events likely caused the extinctions in the oceans at the K-Pg boundary (Robertson et al. 2004; Milne and McKay 1982).

The land food chain is shaped like a pyramid with a small mass of consumers at the top. For instance, where Brian lives in Colorado, mountain lions are at the very top of the natural food chain and occasionally walk through his yard, as do bears. At the base of the food chain is a large mass of food producers, plants. In fact, photosynthesis produces so much food on land that normally there is 10 years of food at the base of the food chain given the current rate of consumption. Of course, it was not normal after the impact. The above-ground biomass was largely destroyed by the fires and then suppressed by several years of bad weather. Undoubtedly, any large creatures that survived the initial impact and fires starved to death. Robertson et al. (2004) noted that the geologic record shows that small creatures that likely lived in holes were favored survivors probably because they still had access to food in the form of roots and insects living in the soil.

Normally food on the land would not be hard to find even if photosynthesis stopped for a year. However, if the global fires destroyed the above-ground biomass, any surviving dinosaurs may have starved. Not only lack of normal plant food but also the difficulty of hunting in the dark would have been a challenge. Many creatures can see at extremely low light levels. For example, people can still see (barely) on a moonless night. Moonless nights have light levels about one-hundred millionth (10^{-8}) of normal. Such low light levels probably only lasted a short time, perhaps a few months after the K-Pg impact (Figure 4.2). Nocturnal creatures, such as many mammals and some birds, would still be able to find food. Creatures used to living underground would still have insects and roots to consume. However, large diurnal animals would find survival difficult because there would be little food left above ground due to the fires, and it would be too dark to easily find any food remaining.

Additional challenges to survival on the land and near the ocean surface would come from rapidly falling temperatures. Figure 4.4 shows that both soot and SO_2 injections lead to temperatures below freezing for several years over the land and to substantial cooling at the ocean surface. Figure 4.5 shows that soot and SO_2 injections also lead to drought conditions in many continental areas, particularly in the tropics. Even if fires had not burned across every forest and grassland, these low temperatures and low rainfall would have prevented new plant growth, further starving large land creatures. In the oceans, many

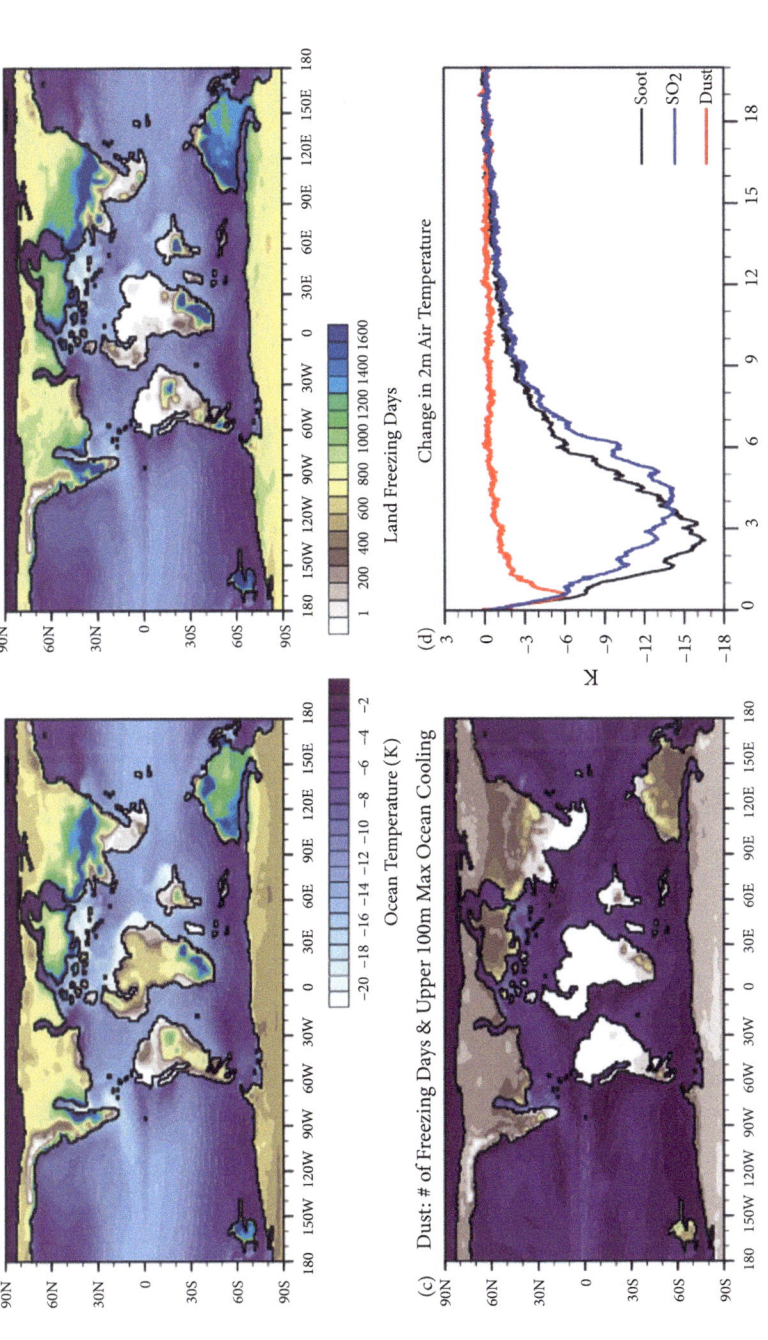

Figure 4.4 Changes in surface temperature following the K-Pg impact due to sulfur dioxide, soot, and dust injections. The temperature falls least due to dust, because the injected mass is so large that the particles coagulate to form large heavy grains that quickly fall out. Soot and sulfur dioxide injections involve smaller masses, so the resulting particles linger, and cold temperatures occur for almost a decade. Figures from Tabor et al. (2020).

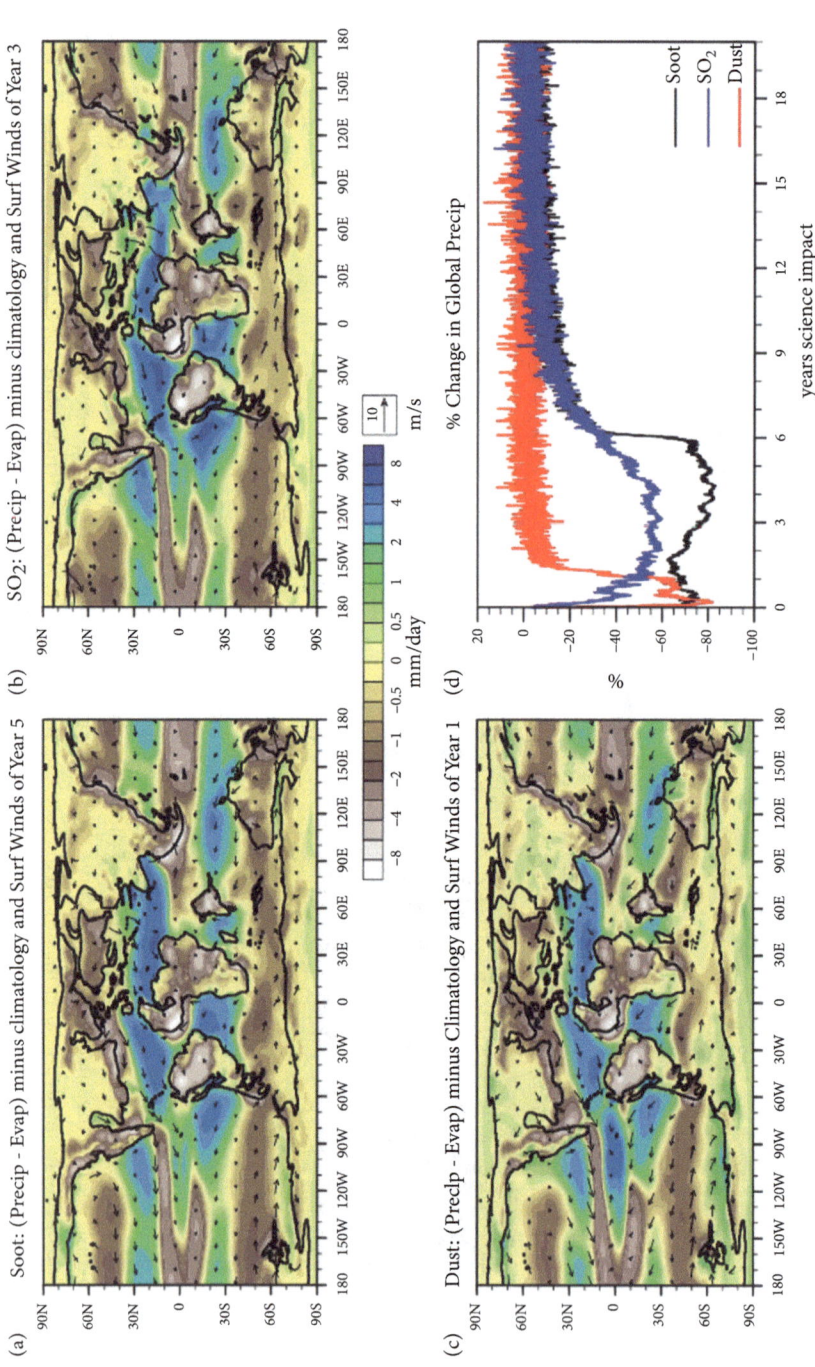

Figure 4.5 Changes in precipitation following the K-Pg impact due to sulfur dioxide, soot, and dust injections. Figures from Tabor et al. (2020).

© 2020. American Geophysical Union

plankton are so sensitive to temperature that their presence or absence is used as a thermometer by scientists.

An alternative suggestion to low light levels triggering the oceanic extinctions is that the oceans became acidic from acid rains. Currently, acid rain is produced from nitrogen oxides generated in engines and power plants when the air is heated and by lightning. Acid rain is also caused when sulfuric acid is produced from the sulfur dioxide generated by burning fossil fuels as well as by volcanic eruptions releasing sulfur dioxide from hot rocks. While it was initially suggested that nitric acid could have caused ocean acidification, it is now thought that not enough nitric acid would be produced when the impact heated the nitrogen in the air to acidify the ocean surface. Likewise, even the release of sulfur from the target rock in Mexico and from the asteroid itself is thought to be insufficient to acidify the ocean surface (Tyrrell et al. 2015).

The Distal layer contains the clues to the causes of the K-Pg mass extinction. Rock dust and sulfur dioxide from both the asteroid and the impact site were injected into the atmosphere during the impact. Some soot may have been produced from combusting hydrocarbons in the target rocks, but most of the soot likely came from forest fires over most of the Earth ignited by the spherules when they entered the atmosphere. Each of these materials is found in the Distal layer, so we know they were distributed worldwide. There is uncertainty about the relative roles of these materials in the extinction. There are advocates, as well as detractors, for each of them. It is likely that each of them contributed in some way. However, low light levels seem to be needed to explain the oceanic extinctions, and these only arise due to the absorption of sunlight by soot. The extinction of the dinosaurs, or at least a severe blow to their abundance, may have occurred when the fires were started. The glowing skies may have produced fatal burns, and the fires would have been so geographically extensive that it would have been difficult not to have been trapped in them. In the decade following the fires, all three of these materials—sulfates, dust, and soot—would have produced subfreezing surface temperatures by blocking sunlight from reaching the surface, destroying plants and eliminating food for land creatures that depend on grazing. Nevertheless, there were refugia from the fires for small land creatures that lived in holes or cavities or in swamps, rivers and lakes. These same dwelling locations and possible nocturnal lifestyles would have allowed creatures to find food, perhaps largely underground for mammals, and survive a prolonged cold spell. In the oceans, life would have been difficult in the photic zone both due to loss of food derived from photosynthesis and due to significant temperature declines. In contrast, life in benthic zones may have been much like normal, and plankton with resting stages and other means for survival during the polar night would have been favored.

An important question to answer is whether or not we could prevent a large asteroid or comet from hitting Earth in the future.

Chapter 5
Can We Stop an Asteroid or Comet Collision in the Future?

5.1 What Size Objects Do We Need to Stop from Hitting Us?

The extinction of the dinosaurs took place long ago, before people even existed, which may delude us into thinking that Earth is seldom hit by objects from outer space. However, that is not the case. As illustrated in Figure 2.1, there are millions of objects with diameters between 20 and 50 m and above whose orbits carry them near Earth. These could destroy a large city if there were a direct hit, and in fact there have been two close calls in the past 100 years, at Tunguska and Chelyabinsk, both of which are in Russia. There are more than a hundred thousand objects whose orbits carry them near Earth that have diameters larger than 100 m and energies larger than 100 Mt, which is a greater energy than that of the largest nuclear weapon ever exploded. There are about 1000 asteroids larger than 1 km in diameter whose orbits cross Earth's orbit, and these objects have energies that exceed 10 times that of Earth's arsenals of deployed strategic nuclear weapons. Objects with sizes near 1 km may be able to create global climate perturbations and perhaps even low-level extinctions. There are probably even a few objects left out there that could collide with Earth as large as the one that killed the dinosaurs that have energies exceeding 100 million of the largest nuclear weapons in the world's arsenals. Objects with sizes from 5 to 10 km might start to trigger mass extinctions. We observe small asteroids entering the atmosphere frequently, and craters show that even within the history of the genus Homo that kilometer-sized objects have hit the planet. Which of these do we need to protect ourselves from—city killers or makers of extinctions—and how could we protect ourselves? The question is open to debate, as we discussed in Chapter 2.

There are two classes of objects of concern with respect to stopping a collision. First, there are objects that have been observed well enough to determine an accurate orbit and that have orbits that do not threaten us for decades or centuries. Given time to develop a strategy, we may be able to stop at least those objects smaller than a kilometer or so. Second, there are asteroids and comets that have just been discovered whose orbits may not be known well enough to

Earth in Flames. Owen Brian Toon and Alan Robock, Oxford University Press. © Oxford University Press (2025). DOI: 10.1093/9780197799734.003.0005

predict a collision or which are likely to hit us within a few years. These objects may be difficult to stop from hitting us.

The orbits of the well-observed objects have been predicted hundreds of years into the future and an impact probability assigned by NASA's Center for Near Earth Object Studies (CNEOS) through its Sentry System (https://cneos.jpl.nasa.gov). Of course, there are uncertainties in the orbits even for well-observed objects. For example, the 0.49 km diameter 101955 Bennu asteroid, which was observed 554 times between 1999 and 2020, is expected to have a probability of 1 in 1700 of hitting Earth between 2178 and 2290. At the current time, the orbit error for the multiple asteroid passes 50 to 150 years from now is very small; however, as we get closer to the time of possible impact, the orbit will be even better known, and the odds of impact will either grow or diminish.

Scout, another program of the CNEOS, reports on recently detected objects. The properties of these objects are not well known, so their orbits are highly uncertain. In fact, some might not even exist and could just be some sort of detection error. The Chelyabinsk object was not detected in advance of its impact by the NASA networks, possibly because it came from the direction of the sun, where astronomers have difficulty observing due to glare of the sun.

5.2 How Can We Prevent an Impact?

Numerous ways have been suggested to prevent an asteroid or comet from hitting Earth. Generally, asteroids have orbits within the orbit of Jupiter, so astronomers can see them easily and have many opportunities to find them and compute an accurate orbit for them. By 2022, NASA's CNEOS surveys had found 117 near-Earth comets and over 28,000 near-Earth asteroids, about 10,000 of which are larger than 140 m in diameter, and 887 larger than 1 km in diameter. The CNEOS survey ranks 2256 of these objects as potentially hazardous because they are larger than 140 m and will pass within 7.5 million kilometers of Earth (about 18% of the nearest approach distance between Venus and Earth). Of these hazardous objects, 158 are larger than 1 kilometer in diameter. While small asteroids may not be detected until they are close to Earth, many larger asteroids may be found years or even centuries before they hit Earth.

We know little about the strengths of asteroid or comets; some may be piles of rubble held together by gravity, while some might be strong solid bodies. We would have to send a spacecraft to each one in advance to determine how to best prevent an impact. It may be possible to alter the orbits of physically strong asteroids by slowly pushing them. Earth moves along its orbit by a distance equal to its own diameter in about 7 minutes, so altering the orbit of an asteroid or comet so that it passes 7 minutes earlier or later would prevent an impact. Ahrens and

Harris (1992) concluded that one could avoid a collision by giving an asteroid a push about a decade before impact to increase its speed by 1 cm per second. In contrast, you'd need to increase the asteroid speed by about 1 m per second if you only got to the asteroid six months before impact.

Melosh and Nemchinov (1993) suggested putting a solar reflector near a dangerous asteroid or comet nucleus discovered far in advance of impact and focusing sunlight on it to slowly evaporate material. The push from the evaporating material would slowly propel the asteroid away from its computed rendezvous with Earth. There are many other ways to push an asteroid. For example, a spacecraft parked near a small asteroid could slowly pull on the asteroid by gravity and move it away from collision. This approach, called a gravity tractor, has the advantage that if the asteroid is structurally weak it might stay together despite the application of a small force. Or one could stand off from the asteroid with a powerful laser and vaporize the rock to create jets to move the asteroid or comet. Or a neutron bomb, which has less blast energy than normal atomic bombs but instead emits a lot of neutrons, might vaporize parts of the rock without threatening to break it into pieces. Or a kinetic impactor, such as a spacecraft, slamming into the asteroid could change its orbit or blast it into pieces whose velocities would take the pieces away from Earth. While each of these ideas might work, none of them has yet been attempted on an asteroid heading toward Earth. But there have been experiments conducted with spacecraft landing on asteroids, spacecraft hurling objects at asteroids, and spacecraft hitting asteroids. Given a century of warning, we could almost certainly devise means to move at least moderate-sized asteroids.

During 2022, in a test of the kinetic impactor idea, a spacecraft named DART (Double Asteroid Redirection Test), slammed into Dimorphos, a small asteroid with a roughly 165 m diameter, in orbit around Didymos, a larger asteroid (Rivkin et al. 2021). The goal was to determine if the orbital period of Dimorphos around Didymos, which is about 12 hours, could be changed by at least 73 s. The experiment planned to determine the efficiency of transferring momentum during the impact. The collision was successful. Dimorphos turned out to be a rubble pile, held together by gravity, and at least 1 million kilograms of ejecta were observed. Dimorphos's orbit slowed by about 33 minutes. This slowing is equivalent to an orbital speed change of 2.7 ± 0.1 mm s^{-1}. The amount of momentum transferred to the asteroid was about 3.6 times greater than that of the impactor due to the large mass of debris flying off the asteroid (Cheng et al. 2023). The success of DART indicates that kinetic impactors can successfully modify orbits of asteroids.

A significant challenge is to stop a large comet moving at high speed and discovered only a short time before impact, or an asteroid or comet that is not detected until the last moment before impact. Comets typically have much higher velocities than asteroids. It can be difficult to determine their size and predict their orbits. Comets are mostly made of various ices and dust. As they draw close to the sun, the ices vaporize and shroud the nucleus, the solid body of the comet, with a halo, called the coma, making it difficult to tell how large the comet is. The coma can be very large compared with the comet core, or it can be quite small if little ice is left. Eventually the pressure of sunlight as well as the solar wind will blow the vapor and dust behind the comet to form one or more tails. The evaporating ices provide some thrust, and the evaporation causes the mass of the comet to change with time; both will change the orbit. Of course, as comets slowly orbit the sun, they lose more and more ice, making them smaller and harder to see.

There are several groups of comets. Halley's comet is the model for about 75 known comets with periods of less than 200 years. Its orbit extends inward between Mercury and Venus and out as far as Pluto. This type of comet can have an orbit that is highly inclined to the plane occupied by the solar system planets. About 511 short-period Jupiter-family comets have orbital periods around 20 years in the plane of the planets and are generally well observed, although their orbits are not predictable very far in advance.

The most dangerous comets are those that fall into the inner solar system from the Kuiper belt or the Oort cloud, both of which are thought to contain large numbers of comets that are far beyond the orbit of Pluto. Their great distance makes it unlikely that modest-size comets would be detected until they entered the inner solar system. Comet Bernardinelli-Bernstein is one of the largest comets ever seen; possibly its nucleus is larger than 120 km. It is thought to orbit the Sun every 3.5 million years, so humans have never seen it before. It was discovered in 2014 and predicted to reach as close into the solar system as the orbit of Jupiter in 2031. Had it been on a course to impact Earth, we would have had only 17 years to decide what to do. We probably would not be able to deflect such a large object on such a short timescale. However, we might survive the impact, as our ancestors did at the time of the K-Pg event, by moving some tiny fraction of humanity far underground with the food necessary to survive for a decade or more and the tools to repopulate the devastated Earth. The energy of this impact could heat Earth's atmosphere far beyond the boiling point of water, leaving little alive on the surface, and boil the upper ocean.

We might not detect a smaller comet until it was only six months or a year from impact, especially if it were hidden because it came from the direction of the Sun. A cautionary tale is the 5 km diameter comet NEOWISE C/2020 F3. It was discovered only four months before making its closest approach to Earth, despite eventually becoming a comet that could be seen with the naked eye. Fortunately, the closest approach was 140 times the lunar distance from Earth; however, an impact could have been disastrous.

Sleep et al. (1989) showed that impacts by asteroids with diameters of several hundred kilometers likely occurred in the early history of the Earth, and the huge amounts of energy deposited during these impacts may have boiled the oceans. Such events may explain why the most primitive organisms on Earth, presumably reflecting the earliest life on Earth, are thermophiles—organisms able to tolerate very high temperatures. Even the K-Pg asteroid had enough energy to ignite fires across the Earth and raise the air temperature at the surface of the Earth due the energy deposited by the impact by many tens of degrees Celsius for several days and evaporate about 10 cm of water from Earth's oceans according to Segura et al. (2013). The Chicxulub crater is the third largest known on Earth. It is estimated that the objects that created the Vredefort (~300 km, 2.02 billion years ago impact date, 20 km diameter asteroid) and Sudbury (~250 km, 1.85 billion years ago impact date, 16 km asteroid) craters could have heated the topmost 5 m of the seas to 50°C, and the lower atmosphere above 100°C for about two weeks with temperatures as high as 600°C for a day or so (Segura et al. 2013). Clearly for human survival, asteroids or comets much larger than Chicxulub cannot be allowed to hit Earth. A large comet impact such as Bernardinelli-Bernstein would likely kill all exposed life on the surface. But could we stop it?

Large asteroids and comets or smaller objects found just before an impact require a large amount of energy be used to prevent disaster. We can evaluate various strategies for dealing with different-sized objects moving at different speeds. There are basically two approaches. First, one can reduce the size of the impactor so that the individual pieces are stopped in the atmosphere instead of hitting the surface. Alternatively, one can deflect the asteroid, or pieces of it, so that they do not hit Earth.

One could try to vaporize the asteroid or comet so that any tiny remaining dust grains are ablated in Earth's atmosphere. This would not prevent an object initially larger than Chicxulub from heating the atmosphere to a high temperature because the vaporized asteroid would have the same impact energy as the original, but it might prevent a smaller body from hitting the surface. There is about 1,000 Mt of energy in the world's nuclear arsenals. Lubin and Cohen (2022) estimate it would take 50 times this much energy to vaporize a 10 km diameter asteroid. So, we cannot protect ourselves from large impactors by vaporizing them with our current arsenals. We might be able to vaporize an asteroid a few kilometers in diameter, perhaps, in the unlikely case that we could hit it with all the world's nuclear weapons simultaneously.

Fortunately, one does not need to vaporize the asteroid or comet. Some asteroids are thought to be rubble piles, in which smaller pieces are held together by gravity. Such asteroids are said to be cohesionless because there are no binding forces between the pieces except gravity. Even a relatively strong nickel-iron asteroid might have fractures or zones of weakness. Philip Lubin has argued that for objects in the size range of 20 to 1000 m, which are not detected in time to alter their orbits, the objects could be fragmented into pieces (Lubin 2023). These small fragments, hopefully smaller than 10 m, would then enter Earth's atmosphere and be destroyed by shock waves produced by drag before hitting the surface, or, if time allowed, the fragments would simply fly out into space and largely miss Earth. Lubin estimates that a 50 m diameter asteroid intercepted five hours before hitting Earth could be fragmented and its pieces end up largely missing Earth if they were given sufficient velocity. The distance from the moon to Earth is about 385,000 km, so an asteroid traveling 20 km/s could travel from the moon to Earth in about five hours, and a comet traveling 50 km/s could reach Earth in two hours. Therefore, interception would need to take place well beyond the moon's orbit and far from Earth, which is challenging from the perspective of the travel time to reach the object with the kinetic impactor. To fragment a 100 m diameter asteroid traveling 20 km/s, Lubin estimates that a kinetic impactor with a mass of a few kilograms that transfers momentum efficiently with a fragile asteroid, as DART did, could fracture the asteroid. However, energy coupling to the asteroid interior would be an issue, and a much larger impactor would likely be required.

Arakawa et al. (2020) describe hitting the 1 km in diameter asteroid (162173) Ryugu from the Hayabusa2 spacecraft with a metal ball weighing 2 kg and traveling at 2 km/s. The surface of the asteroid has boulders up to 160 m in diameter lying on a regolith (a layer of fine rock grains likely created by bombardment by smaller objects) that was expected to be held under gravity by cohesion forces between regolith grains. The impact caused a crater with a diameter of 14.5 ± 0.8 m and a depth of about 2.2 m. The ejected material largely fell back onto the asteroid near the crater, which suggested to Arakawa et al. that the regolith was effectively cohesionless, like dry sand. The asteroid didn't fracture.

A'Hearn et al. (2005) discussed the 2005 impact of one of the Deep Impact spacecrafts with comet Tempel 1, a typical member of Jupiter's family of comets, with a mean radius of about 3.0 km. The impacting spacecraft and its residual fuel weighed about 370 kg and was traveling at 10 km/s. A large volume of ejecta material fell back onto the comet, suggesting it was held together by gravity and was "an extremely weak, powderlike substance." Based on Earth-based observations of the mass of material ejected by the impact, the crater has a diameter of about 130–220 m and may have excavated into a layer of the comet 10–30 m deep, with different composition than the surface. The comet didn't fracture.

These observations of asteroid and a comet cratering by kinetic impactors indicate some asteroids and comets could be composed of sand-like material, at least on their surfaces, which can be easily disintegrated. However, using an impactor as small as a few kilograms to dismantle a 100 m diameter asteroid as suggested by Lubin would be highly risky given our limited information about the ability of impacts to destroy relatively weak objects that are large and given the lack of knowledge we would have of the cohesion of a newly discovered asteroid or comet. Ahrens and Harris (1992) suggested that the orbit of a 100 m diameter asteroid could be altered by 1 cm/s by hitting it 10 years before impact with a 200 kg projectile traveling at 12 km/s. Presumably, a 100 times more massive projectile would be needed if only six months were available to prevent it from hitting Earth. Ahrens and Harris (1992) suggested that the orbit of a 10 km diameter asteroid with a cohesionless structure could be altered with six months of lead time using an explosive energy near or above 1000 Mt in the form of neutron bombs. The bombs would boil away the regolith, adding an impulse to the asteroid. Such weapons do not exist currently.

Lubin and Cohen considered the seemingly impossible task of stopping a 10 km diameter asteroid collision at 40 km/s when the asteroid is discovered six months before impact, leaving just enough time to launch a spacecraft five months before impact. They find that just the transit time for the spacecraft traveling 10 km/s to reach the asteroid leaves only a month before impact to stop a collision. Once the spacecraft reaches the asteroid, they suggest that 10 Mt of energy, produced from nuclear explosions, would be enough to fragment the asteroid if it were held together by gravity and to give the pieces a velocity of 5 m/s. If the fragments were traveling transverse to the impactor's orbit, they would spread out to about twice the diameter of Earth for 30 days and therefore mostly miss Earth. They envision that a single large bomb would disintegrate on impact rather than penetrating the asteroid and that it would be better to use smaller penetrator weapons to "onion peel" the asteroid. In this case, the first explosions would peel away the outer layers of the asteroid, until interior portions could be reached by later explosions to fragment the asteroid.

While numerous ideas have been developed to prevent an impact, none of these have been carried far enough to know that they could be relied on. Extensive engineering analyses need to be carried out, and eventually we will need to practice as was done by DART. We also need more information about the properties of asteroids and comet nuclei. Continued discovery of asteroids and comets, their sizes, and their orbits are also critical. Discovering objects decades or centuries before impact would make it much more practical to prevent a collision.

Could we protect ourselves now from an object such as the 120 km diameter comet Bernardinelli-Bernstein discovered 17 years before closest approach, or the 5 km comet NEOWISE C/2020 F3 discovered 4 months before closest approach? Almost certainly not. These would likely require nuclear weapons to prevent an impact, and the numbers and yields of such weapons do not currently exist and neither do the delivery systems. Very large energies, implying multiple coordinated nuclear explosions, would likely be needed. We lack the plans to prevent such impacts. While heavy launch vehicles such as the SpaceX Starship or the NASA Space Launch System would probably be needed to carry explosives or other means to change the orbit of the asteroid or comet, these are not yet operational but should be in the next few years.

Possibly in coming decades we will understand more clearly what we need to do to defend our planet. In the meantime, much like preventing nuclear wars, we have to depend on luck to prevent global disaster. Impacts may be more predictable than nuclear wars, at least for the asteroids we can detect (because they obey physics, unlike human behavior). However, comets could impact with little warning since many of them reside so far away that we can't see them with current technology. No large objects are known that currently threaten another extinction event such as the one that occurred 66 million years ago.

An important question to answer is whether we could survive a mass extinction event caused by an asteroid the size of the one that killed the non-avian dinosaurs. The dinosaurs themselves arose after a mass extinction, and then they became the dominant large land creatures following another mass extinction event, only to fail on a third exposure.

Humans, like our distant small mammal ancestors that survived the K-Pg mass extinction, tend to live in caves, called houses, and visit other types of shelters, such as concrete parking structures, some of which could protect us from the initial glowing skies of a swarm of shooting stars produced by an asteroid impact. In subsequent days, global fires would likely destroy most of our cities and other dwellings and kill a large fraction of the population. Years of darkness and cold temperatures would halt agriculture globally, destroy marine food sources, and doom most of us to starvation. Destruction of infrastructure would make heating our remaining dwellings difficult, we would soon run out of fuel for cars, and communication would break down. Perhaps a few of us would survive, fortunate enough to have made it through the initial deadly threats from fires and lucky enough to find stored food and ways to stay warm.

No large asteroids or comets are aimed at us now, as far as we can tell. However, there are nuclear weapons aimed at us. Nuclear wars can produce some of the same effects as an asteroid impact. Now let's turn our attention to whether we can survive, or better yet prevent, a nuclear war.

PART II
HUMANS AND NUCLEAR WINTER

Chapter 6
You Too Could Build a Bomb

It Can't Be Hard; There Are a Lot of Them

6.1 Where Does the Energy Come From?

Nuclear weapons produce energy by converting mass into energy either by fission, in which a heavy element splits apart and creates lighter elements, or by fusion, in which a low-mass element fuses with another low-mass element to create a heavier element. Surprisingly, the energy of nuclear weapons and of asteroid collisions both originate from the formation of the solar system. Nuclear elements heavier than iron, such as uranium, were created before the Sun was born and released into a stellar nursery by the explosions of supernovas and red giants, the remnants of dying stars, and the collisions of neutron stars. Of course, most of the elements in our bodies were formed in stars, so we too are made of star stuff.

The power of nuclear weapons is described by the famous equation discovered by Albert Einstein, $E = \Delta m \, c^2$. This discovery was shocking. It is hard to imagine that just living, or kicking a ball, results in a loss of mass, Δm. Energy and mass are different forms of the same thing, and whenever energy is produced, mass is lost. We can rewrite this relationship as $\Delta m = E/c^2$. This expression means that in any physical or chemical process that releases energy an amount Δm of mass is lost, which equals the energy released divided by the speed of light, c, squared. Even normal chemical reactions follow this relationship, and anything you do, such as kicking a ball or breathing, results in a mass change. These mass changes are so small they cannot be measured. But nuclear reactions release a lot of energy, so the mass changes can be determined. The speed of light is a very large number, about 300 million meters per second (671 million miles per hour). The speed of light is fast enough for light to travel to a geostationary satellite parked 36,000 km (about 22,000 miles) over Earth's equator and back to the ground at a latitude of 45 degrees in about one-quarter second. This delay is noticeable when you hear news reporters talking to someone in a distant location since it requires double bounces of the light beam from the studio news reporter to the satellite then down to the reporter in the field to ask a question and then another double bounce for the distant reporter to respond with their answer. Of course, some of the delay is due to the reporter in the field thinking

Earth in Flames. Owen Brian Toon and Alan Robock, Oxford University Press. © Oxford University Press (2025).
DOI: 10.1093/9780197799734.003.0006

about their response. Since c is the fastest possible speed of any object, no object can have a kinetic energy, $0.5\ m\ v^2$, greater than the energy contained in the mass of the object.

Atoms are composed of electrons, protons, and neutrons. Electrons are relatively low-mass objects surrounding the atomic nucleus in a cloud. Chemical reactions involve the electrons in atoms and molecules. The nucleus is composed of neutrons and protons. Nuclear reactions involve these components of the atomic nucleus.

Figure 6.1 shows the typical mass of protons or neutrons, which together are called nucleons if they are in the nucleus of an atom, for various elements. The nucleons of iron have the lowest mass per nucleon. The mass of helium nucleons is less than the mass of hydrogen nucleons. Therefore, if hydrogen is converted to helium by nuclear fusion, energy is released in an amount that equals the change in mass times the speed of light squared. Similarly, if uranium is converted to lower atomic mass elements by nuclear fission, the mass per nucleon declines, and energy is released.

The simplest uranium fission weapons were not very efficient. For example, the bomb dropped on the Japanese city of Hiroshima during World War II

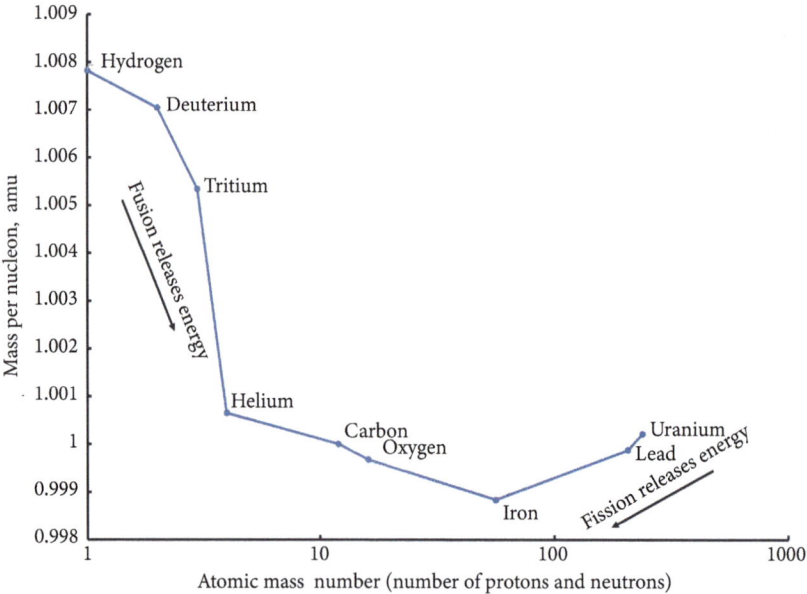

Figure 6.1 The mass per nucleon (a neutron or proton) as a function of atomic mass number of the elements (total number of protons + neutrons in the element). The mass per nucleon is measured in atomic mass units, amu. An amu is one-twelfth of the mass of a single atom of carbon, which has an atomic mass of 12 and 12 nucleons. 1 amu is 1.66×10^{-24} g.

contained about 64 kg (140 pounds) of uranium. The uranium was enriched to contain 80% of the isotope ^{235}U, but less than a kilogram (2 pounds) underwent fission, and only about 0.63 g (1/50 of an ounce), about the weight of a large bumblebee, was converted into the 15,000-ton (15 kiloton) energy release. The weight used to measure the energy of the bomb is the weight of the amount of conventional explosive that would release the same amount of energy as the atomic bomb and is not related to the actual physical weight of the bomb. The total weight of the bomb dropped on Hiroshima was about 4 tons. Similarly, a mass of about 4 million tons is lost from the Sun every second during nuclear fusion, which powers the Sun's energy loss mainly in the form of visible light. Four million tons is about four times the weight of the heaviest object ever transported from one place on Earth to another, a natural gas platform offshore from Norway. Don't worry; over its remaining lifetime of about 5 billion years from now (the Sun is middle-aged since it is now about 5 billion years old), the Sun will only lose a few hundredths of a percent of its mass due to fusion.

Figure 6.2 shows a model of the Hiroshima bomb and the actual airplane that dropped that bomb. This plane, the Enola Gay, named after the mother of the pilot, Paul Tibbets, has been refurbished and is now on display at the

Figure 6.2 Author Alan Robock in front of the Enola Gay, the airplane that dropped the bomb on Hiroshima on August 6, 1945. The front half of the fuselage of the airplane is behind him, and the rear half is to the left. They have now been reassembled and the plane is on display at the Smithsonian Air and Space Museum Steven F. Udvar-Hazy Center near Dulles Airport in Virginia. A model of the "Little Boy" uranium bomb that was dropped on Hiroshima is in front of him.

Udvar-Hazy Center of the Smithsonian Air and Space Museum, near Dulles Airport in Virginia.

Nuclear weapons don't release all their energy at once because the radioactive elements produced in the fission process decay over varying periods of time. In some cases, this decay occurs over many years, resulting in the release of additional energy. Usually only the energy released in a minute or so after the explosion is included in the equivalent tonnage of the nuclear weapon yield. Nevertheless, this unit of a ton, which does not correspond in any way with the weight of a nuclear weapon, provides a visual comparison with conventional bombs that might weigh close to the amount corresponding to their actual explosive yield, since conventional bombs and other explosives are often based on TNT. A kiloton of TNT would occupy a cube about 8.5 m (28 feet) on a side, much too large for any aircraft to carry. TNT is a very safe liquid to work with and can be poured into bomb casings, but TNT is not the most efficient explosive. For example, dynamite, which was invented by Alfred Nobel, releases more energy per kilogram of explosive. It is a mixture of nitroglycerine, which is very unstable and explosive, with clay to make it safe to handle. Nobel thought that the horror of using dynamite as a weapon would be so terrible that it would help end war. When he found out that he was sadly mistaken, he used the money he had earned from it to endow the Nobel Peace Prize, awarded to "the person who has done the most or best to advance fellowship among nations, the abolition or reduction of standing armies, and the establishment and promotion of peace congresses."

The conventional bombs exploded in World War II, dropped by fleets of hundreds of aircraft on numerous missions over many years, are estimated to have released an energy of about 3 million tons (megatons or Mt) of TNT. A single nuclear armed aircraft can carry nuclear bombs with the same explosive power as all the bombs dropped in WWII. The average typhoon or hurricane releases about 10 Mt of energy per minute, while the Earth receives about 62 Mt per minute of sunlight. The almost 60,000 wildfires in the United States in 2020 released about 100,000 Mt of energy. As you can see, the energy released from natural processes dwarfs that controlled by humans.

6.2 Nuclear Reactions: Fission and Fusion

The energy from nuclear explosions comes from the conversion of mass into energy. Most energy production we are familiar with comes from chemical reactions, or other interactions, that involve the electrons in atoms. There is so little energy released in these that the mass changes are too small to be directly measured at present. However, nuclear reactions involve the protons and neutrons that make up the nucleus of atoms. In this case, the difference in mass between the atom and its fission daughter products can be measured. Figure 6.3 provides an example of the mass change when a hydrogen atom of mass number 2 (called deuterium) is assembled from a proton, a neutron, and an electron. The mass of the components is slightly greater than the mass of the deuterium atom, so that energy is released when deuterium is made.

While the equivalence of mass and energy, $E = m\,c^2$, is the famous way to think about nuclear reactions, other ways are more intuitive based on our normal experiences. The protons in the atom's core are electrically charged. The atomic number of an element is the number of protons in an atom of the element. Uranium has 92 protons. If an atom is uncharged, the atomic number is also the number of electrons.

The number of neutrons in an atom of an element is not fixed, and atoms with different values are called isotopes. They are tracked by their atomic weight. There are at least 27 isotopes of uranium, but the naturally occurring isotopes are ^{238}U (99.28%), ^{235}U (0.71%), and ^{234}U (0.0054%). The number in parentheses is the fraction of natural uranium made of that isotope, and the superscript appearing before the U is called the mass number, which is approximately the atomic mass of the isotope in atomic mass units, amu. Sometimes there is also a

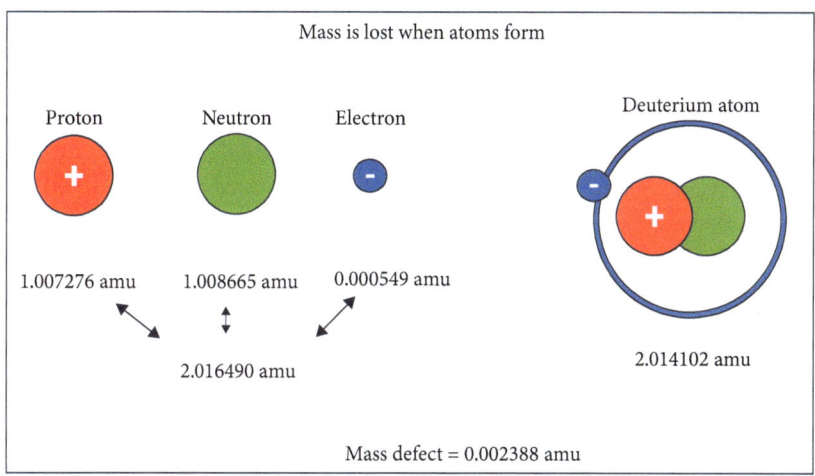

Figure 6.3 The figure presents the masses of isolated protons, neutrons, and electrons, and the mass of deuterium, an isotope of hydrogen, which consists of one neutron, one proton, and one electron. The mass lost when deuterium is formed is called the mass defect. The mass defect is converted to energy when deuterium is formed. Atoms are mostly empty space in the sense that the volume of the hydrogen nucleus is about one-trillionth percent of the volume of the atom. Electrons are not located in a single place but simply have a probability of being found within the atom. Electrons don't amount to much; the mass of the nucleus is more than 99.9% of the mass of the atom.

subscript on the left of the element abbreviation, such as 92 for uranium, which is the atomic number. The mass number is the sum of the protons and neutrons in the isotope. Therefore, $^{235}_{92}$U has 92 protons and 143 neutrons.

All three of the uranium isotopes are radioactive. A radioactive material spontaneously disintegrates, releasing fast-moving particles and/or energetic light rays that can be dangerous to exposed people. The half-life of a material is the time that it takes half of the material to disintegrate. As a rule, after 7 half-lives only about 1% of the original material will be left. The half-life of ^{238}U is 4.5 billion years, that of ^{235}U is 700 million years, and that of ^{234}U is 250 thousand years. ^{238}U is the primary source of natural radiation on Earth, as it disintegrates into a series of lighter atoms. ^{235}U is the only fissionable natural isotope that is primordial and found in nature in abundance. Fission is the process of a larger atom disintegrating into lighter atoms. Fission can happen spontaneously, but also by neutrons hitting the atom forcing it to disintegrate. Given its short lifetime compared with the lifetime of Earth, ^{235}U was once much more abundant, and in fact there is some evidence for nuclear fission occurring in some ancient uranium deposits.

Sidebar 6.1

The nuclear strong force is one of the four true forces in Nature along with gravity, electromagnetism, and the nuclear weak force. Protons and neutrons are composed of several types of quarks and gluons, which are thought not to be divisible into even smaller objects. The strong force binds together the quarks that compose neutrons and protons, and also binds together the protons and neutrons. The strong force is the strongest force in Nature at very small distances. However, the strong force declines rapidly with distance, so the electrical repulsion of protons in large atoms can overcome the strong force.

If protons and neutrons were simply tightly packed in the atomic core, they would fly apart due to electrical repulsion of the protons. Indeed, that is what happens to provide the energy from nuclear explosions, as illustrated in Figure 6.4 for one of many possible nuclear reactions. A slowly moving neutron hitting ^{235}U and joining its nucleus creates an unstable atom of ^{236}U. That atom then splits into an atom of barium (^{144}Ba) and an atom of krypton (^{89}Kr). The mass of the products is less than the original mass by a few tenths of a percent. The missing mass is converted into energy. The products go flying apart at a few percent of the speed of light, releasing an energy hundreds of millions of times

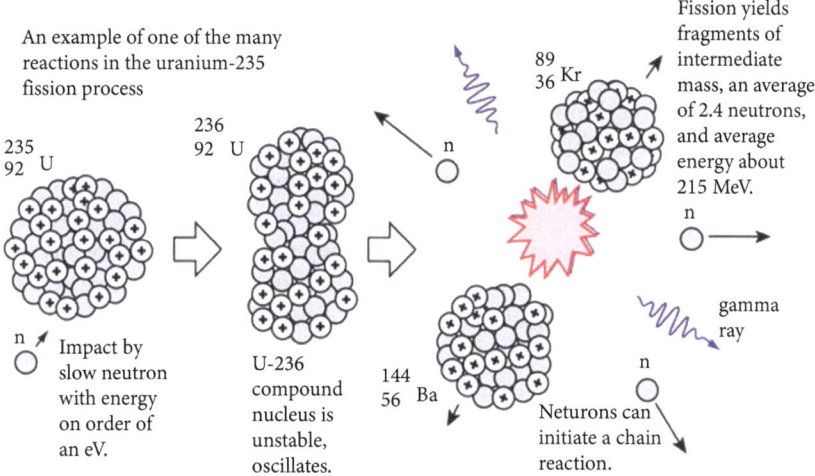

Figure 6.4 One of many possible fission processes in ^{235}U. n is a neutron. 1 eV is 1 electron volt, a unit of energy equivalent to 1.6×10^{-19} joules. 215 MeV (million electron volts) is 3.4×10^{-11} joules, but this energy release is just per atom of uranium.

© HyperPhysics by Rod Nave, Georgia State University. http://hyperphysics.phy-astr.gsu.edu/hbase/NucEne/U235chn.html. Used by permission

greater than the energy of the neutron that split the atom of ^{235}U. Also released are three neutrons and some energetic light in the form of gamma rays. About 85% of the energy released is in the motions of the krypton and barium fission fragments. These interact strongly with their surroundings, and so their energy is quickly converted into heat. The heat forms the fireball of a nuclear explosion, which produces a flash of intense light, or thermal radiation, which can burn people and set fires. The heat in the fireball also generates shock waves that can knock down buildings. The light and the shock waves carry roughly equal amounts of energy. The gamma rays and the neutrons can travel long distances and can, along with electrons and other particles, form initial ionizing radiation that lasts about one minute after a nuclear explosion and contains around 5% of the energy released.

Ionizing radiation can knock electrons off of atoms, making them into electrically charged particles called ions. This nuclear radiation can burn exposed people nearly instantly in a small area near the explosion site. The residual nuclear radiation that remains after about one minute, which people often think about as the great danger of nuclear weapons, is caused by the many different fission fragments, some of which are unstable and disintegrate with half-lives of days to many years, releasing harmful high-energy particles and light. The energy in the residual radiation is 5 to 10% of the total energy released.

The electrical repulsion of the positively charged protons in the nuclei of atoms usually does not cause the atomic nucleus to fly apart because the neutrons and protons are held together by the nuclear strong force. The attractive nuclear strong force diminishes much more rapidly with distance than the electrical repulsive force. Therefore, as atomic nuclei become larger, the strong force is overwhelmed by the electromagnetic forces between the protons, and the atoms break apart. This disintegration is why there are some artificially created elements with large numbers of protons and neutrons that are only stable for very short periods of time. However, even relatively stable elements such as ^{235}U can disintegrate when the nuclei are hit with a neutron, as in Figure 6.4.

When the core of ^{235}U is hit by a rapidly moving neutron, the uranium disintegrates into elements with smaller atomic numbers. Two or three neutrons traveling at a few percent of the speed of light are also emitted, as shown in Figure 6.4. If these neutrons hit other atoms of ^{235}U, even more neutrons are emitted, which then can hit more ^{235}U atoms, leading to a chain reaction of disintegrations, as illustrated in Figure 6.5. This "chain reaction" is the basis of nuclear fission.

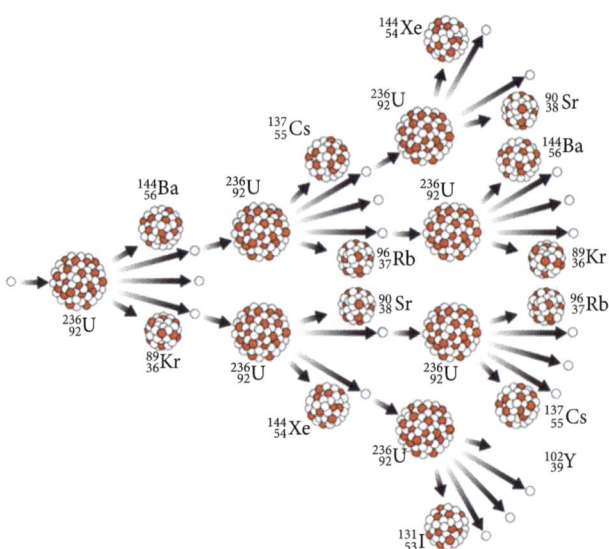

Figure 6.5 In a nuclear chain reaction, the fission of one uranium atom due to being hit by a neutron (white circle) leads to the emission of two or three more neutrons. Some of the neutrons might be lost from the system, but some will hit other uranium atoms, releasing more neutrons. Some fission products will also decay, leading to additional neutrons (not shown).

In practice, the supercritical mass (the minimum needed for an explosion) in the Hiroshima bomb was assembled in about a millisecond in a gun-type bomb. In a gun-type bomb, as shown in Figure 6.6, a conventional explosive is used to fire one part of the critical mass into the other part at high speed. Gun-type weapons are unsafe since an accident, such as an airplane crash, could cause the subcritical masses to combine, and this style of weapon is no longer used by the major nuclear powers. However, due to its simplicity, it might be used by a terrorist group or a country just developing nuclear weapons.

Were you able to obtain weapons-grade ^{235}U, you could create your own bomb relatively easily. Modern weapons-grade uranium is uranium that has been enriched so that about 90% is in the form ^{235}U. Luis Alvarez helped develop the complex mechanics of forming the critical mass in the plutonium atomic bomb described below. He also flew in an observation B-29 aircraft, the Great Artiste, that accompanied the Enola Gay when it dropped the gun-type bomb on Hiroshima. His job was to determine the yield of the explosion by measuring the shock wave. According to Alvarez, you could take two pieces of weapons-grade uranium weighing more than 26 kg each, put one on the ground and drop the other onto to it, and have a good chance that a high-yield explosion would occur. You would not have good control of the energy released, and you would receive a lethal dose of radiation, but you would have a bomb.

Purified ^{235}U is difficult to produce. ^{235}U composes only 0.7% of natural uranium. Nuclear power reactors can usually use low enriched uranium, which is

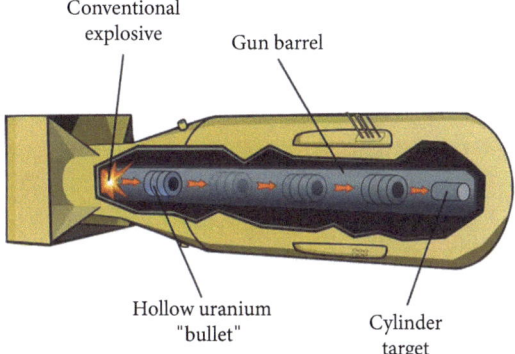

Figure 6.6 A uranium atomic bomb. When the conventional explosion forces the "bullet" and "target" uranium pieces together, it makes a supercritical mass, and the bomb will explode because of nuclear fission. This is the type of bomb that was dropped on Hiroshima on August 6, 1945.

Image from https://upload.wikimedia.org/wikipedia/commons/thumb/b/b7/Gun-type_fission_weapon_en-labels_thin_lines.svg/2560px-Gun-type_fission_weapon_en-labels_thin_lines.svg.png

3–5% ^{235}U. Highly enriched uranium (HEU) is 20% ^{235}U or greater. About 90% of the work needed to produce weapons-grade uranium is completed when 20% ^{235}U is produced. It is estimated by the International Panel on Fissile Materials (2024) that in 2023 there were about 1245 tons of HEU in the world, enough to make about 23,000 Hiroshima-type bombs. Of this HEU, about 1100 tons is in weapons or available for weapons, leaving about 145 tons, enough for about 2800 bombs, in other uses. HEU is used in nuclear propulsion systems on ships, in research reactors, in low power reactors, in some satellites, and to produce medical radioactive isotopes, all of which present a proliferation danger.

^{235}U cannot be chemically separated from the other isotopes. Instead, it is usually separated using the small mass difference between the isotopes or small differences in the spectra of uranium compounds with different isotopes. This separation can be done in several ways, but it is most often done in a centrifuge. The centrifuges in Figure 6.7 are spinning cylinders in which the heavier material, ^{238}U, is forced to the outside by centrifugal force. Centrifugal force is what you experience pushing you outward on a spinning ride. The mass differences between the isotopes are so small that many high-speed centrifuges are required to run over a long time to produce uranium that is highly enriched in ^{235}U.

There are other isotopes that can be used by fission weapons, such as the plutonium isotope ^{239}Pu. ^{239}Pu has such a short half-life (24,100 years) that it is not found in nature. Instead, it is produced in nuclear reactors from ^{238}U. Figure 6.8 shows the original reactor used to produce the plutonium for the bomb dropped in WWII on Nagasaki, Japan. Figure 6.9 shows a model of the Nagasaki bomb. In general use, power reactors produce several isotopes of plutonium, including ^{240}Pu, which contaminates the ^{239}Pu and makes plutonium difficult to use in bombs. However, reactors can be run for short periods of time, which results in

Figure 6.7 A bank of centrifuges to separate uranium isotopes

Image from https://discover.lanl.gov/publications/1663/2020-august/centrifuge-sentries/ and image is https://cdn.lanl.gov/bd728d21-b8ee-47db-b15f-1366d4744d0a.jpg

Figure 6.8 Author Alan Robock in August 2017 at the B-reactor at Hanford, Washington, where the plutonium used in the Trinity test and the Nagasaki bomb was produced. The reactor was placed in an empty region of eastern Washington in case there was an explosion and to isolate people from the pollution. This is the reactor front face, where rods of uranium were inserted as fuel to make the plutonium. Although the reactor also generated heat, none of it was used for other purposes, such as generating electricity.

relatively pure ^{239}Pu (Figure 6.10). The ease of making plutonium in reactors is one reason nuclear reactors are closely monitored to make sure that they are not producing weapons-grade ^{239}Pu by rapidly cycling their fuel.

^{239}Pu has a smaller critical mass, about 11 kg (24 pounds), than ^{235}U, so less plutonium is needed for a bomb. ^{239}Pu cannot be used in gun-type weapons due to the large number of neutrons that contaminating ^{240}Pu may emit. Instead, it is used in implosion weapons, as illustrated in Figure 6.11. In an implosion weapon, a conventional explosive is used to produce a symmetric shock wave that compresses a spherical ball of plutonium, called a pit, so that the critical mass is produced at the center of the ball, leading to a fission explosion. The Nagasaki bomb used 6 kg of ^{239}Pu, which is less than the critical mass because the compression of the pit reduces the mass needed to be supercritical.

It is possible to obtain plutonium by chemically treating power plant nuclear waste to remove other elements such as uranium. A large quantity of nuclear waste is stored near power plants around the world. The reactors in power

Figure 6.9 Author Alan Robock with a model of the "Fat Man" plutonium bomb that was dropped on Nagasaki on August 9, 1945, at the Nagasaki Atomic Bomb Museum, July 2013

plants are usually fueled with low enriched uranium, and only a small fraction of the uranium is destroyed in normal operation. Some reactors are designed to make relatively pure ^{239}Pu for nuclear weapons as the uranium undergoes fission. In normal power plant reactors, the spent fuel is contaminated by heavier plutonium isotopes, particularly ^{240}Pu. The International Panel on Fissile

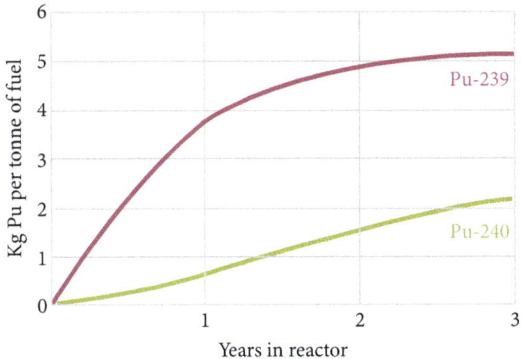

Figure 6.10 An example of the production of ^{239}Pu and ^{240}Pu in a nuclear reactor. The first few days or weeks of running the reactor produces relatively pure ^{239}Pu.

Image from https://world-nuclear.org/information-library/nuclear-fuel-cycle/fuel-recycling/plutonium. Used with permission. World Nuclear Association

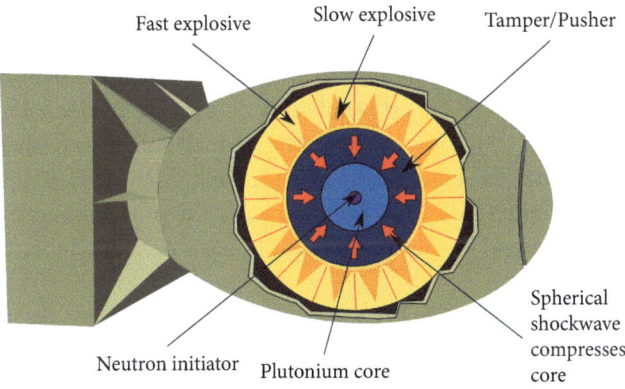

Figure 6.11 A plutonium atomic bomb. This explodes when the plutonium is compressed by the explosives to start the fission chain reaction. This is the type of bomb that was dropped on Nagasaki on August 9, 1945, killing about 60,000 human beings.

From User: Ausis Original png format drawing by User: Fastfission—Original png format drawing, CC BY-SA 2.5, https://commons.wikimedia.org/w/index.php?curid=1549158

Materials estimates there were 540 tons of separated plutonium in the global stockpile in 2023. Organizations outside weapons programs had 420 tons of this plutonium. The global stockpile is enough to produce about 70,000 plutonium nuclear bombs of the type used at Nagasaki. Experts suggest such contaminated fuels could be used to make a bomb just as easily as with enriched ^{239}Pu (Mark 1993). However, it would be technically challenging to make an implosion bomb.

There are two other general types of nuclear weapons. Boosted fission weapons are designed to use a larger fraction of the plutonium than standard fission weapons. The Hiroshima ^{235}U bomb fissioned about 1.4% of its fuel, and the Nagasaki ^{239}Pu bomb fissioned about 13% of its fuel during their explosions. Boosted fission is designed to use the fuel more efficiently not only to get a higher yield but also to reduce the weight of the bomb to make it more practical to carry on missiles. There are many variations of boosted fission weapons, but the basic idea is to supply more neutrons, or to prevent neutrons from escaping, before the explosion disperses the plutonium. Generally, this is done by adding deuterium and tritium as gases inside the fission bomb. When the bomb ignites, the high temperatures create nuclear fusion of the deuterium and tritium, which in turn produces an intense burst of high-speed neutrons, causing more fission fuel in the bomb to explode. By adjusting the amount of deuterium and tritium, and by other methods that basically make the bomb less efficient, one can make variable-yield weapons.

The third type of nuclear weapon is a thermonuclear bomb, often called a hydrogen bomb to differentiate it from an atomic bomb that depends only on fission for its energy. Thermonuclear weapons can produce explosions with megaton yields. The largest bomb ever detonated, the Tsar Bomba, had a yield near 50 Mt, but parts of the bomb, which might have doubled the yield, were omitted to reduce radioactive fallout. Almost all weapons now deployed by the United States, Russia, Britain, France, and China are thermonuclear bombs. India and North Korea claim to have developed thermonuclear bombs.

Thermonuclear bombs work in a similar fashion to boosted fission bombs but introduce an additional second stage of explosion, as illustrated in Figure 6.12. Basically, a fission bomb is placed close to fusion fuel (deuterium, tritium, or lithium deuteride.) When the first or primary stage, a fission bomb, explodes, the fusion fuel in the second stage is compressed and heated to high temperatures. The second stage also has a small inner core of uranium or plutonium that undergoes fission that then drives the fusion, which then drives additional fission in the outer layer of the second stage. In most high-yield weapons, about half of the yield (the so-called fission yield), and nearly all the radioactivity, comes from the fission of uranium and plutonium.

Taking the mass and the velocity into consideration the energy of the asteroid that killed the dinosaurs was roughly 10 billion, 10^{10}, times as powerful as the Hiroshima bomb. We know there are millions of powerful asteroids and comets lurking in deep space. How many bombs are there hiding in countries around the world?

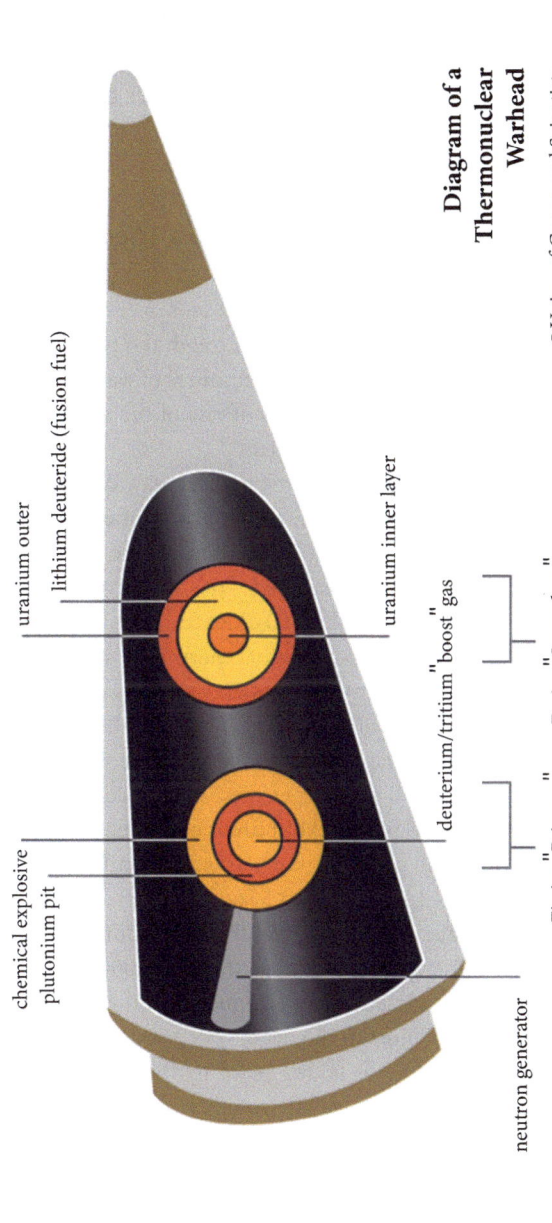

chemical explosive

plutonium pit

uranium outer

lithium deuteride (fusion fuel)

uranium inner layer

deuterium/tritium "boost" gas

Fission "Primary"

Fusion "Secondary"

neutron generator

Diagram of a Thermonuclear Warhead

©Union of Concerned Scientists

Figure 6.12 A hydrogen fusion bomb. These are much more powerful than fission bombs, and theoretically there is no limit on their explosive power. The Soviet Union once tested one with an explosive power of ~50 Mt (50,000 kt). From the Union of Concerned Scientists, used by permission

Chapter 7

How Many Bombs Are Out There, and How Could They Be Delivered?

7.1 How Many Nuclear Weapons Are There?

Figure 7.1 shows the history of nuclear warheads (Kristensen et al. 2024c). From our perspective being more than 20% of the way through the twenty-first century, the graph at first seems to simply record the numbers of warheads through time. In fact, Figure 7.1 charts major landmarks of one of the most important periods in the history of our species on the planet.

Figure 7.1 begins in 1945 at the end of the first nuclear war, WWII, when the United States dropped its two existing nuclear weapons on the civilian population centers of Hiroshima and Nagasaki. These attacks were a continuation of the incendiary bombing raids that devastated cities in Europe, principally Dresden and Hamburg, as well as more than 60 cities in Japan. These raids on civil populations were justified on the grounds that there were military personnel spread out in the cities or industrial facilities and workers contributing to the war, including such things as drill presses in individual homes in Japan. However, General Curtis LeMay, who organized the Japanese raids, said, "We better win this war, or I'll be tried for war crimes." The incendiary raid on Tokyo using "conventional" bombs required hundreds of bombers, of which 8% were lost. The Tokyo raid burned about 41 km^2 (16 square miles); the atomic bomb on Hiroshima burned 11–13 km^2 (about 5 square miles). Fatalities in the Tokyo raid are variously estimated at 75,000 to 200,000. According to Wellerstein (2020), "The United States military estimated that around 70,000 people died at Hiroshima, though later independent estimates argued that the actual number was 140,000 dead. In both cases, most of the deaths occurred on the day of the bombing itself, with nearly all of them taking place by the end of 1945." The atomic bombings only took one aircraft each (two other planes were present to evaluate the results) and none were lost.

At the time, it seemed that the atomic bombings ended the war and avoided an invasion of Japan, which may have led to many more deaths. Brian's father was a naval aviator at that time, flying photo reconnaissance of Japanese-held islands and the Japanese mainland before they were invaded or bombed. Alan's

Earth in Flames. Owen Brian Toon and Alan Robock, Oxford University Press. © Oxford University Press (2025).
DOI: 10.1093/9780197799734.003.0007

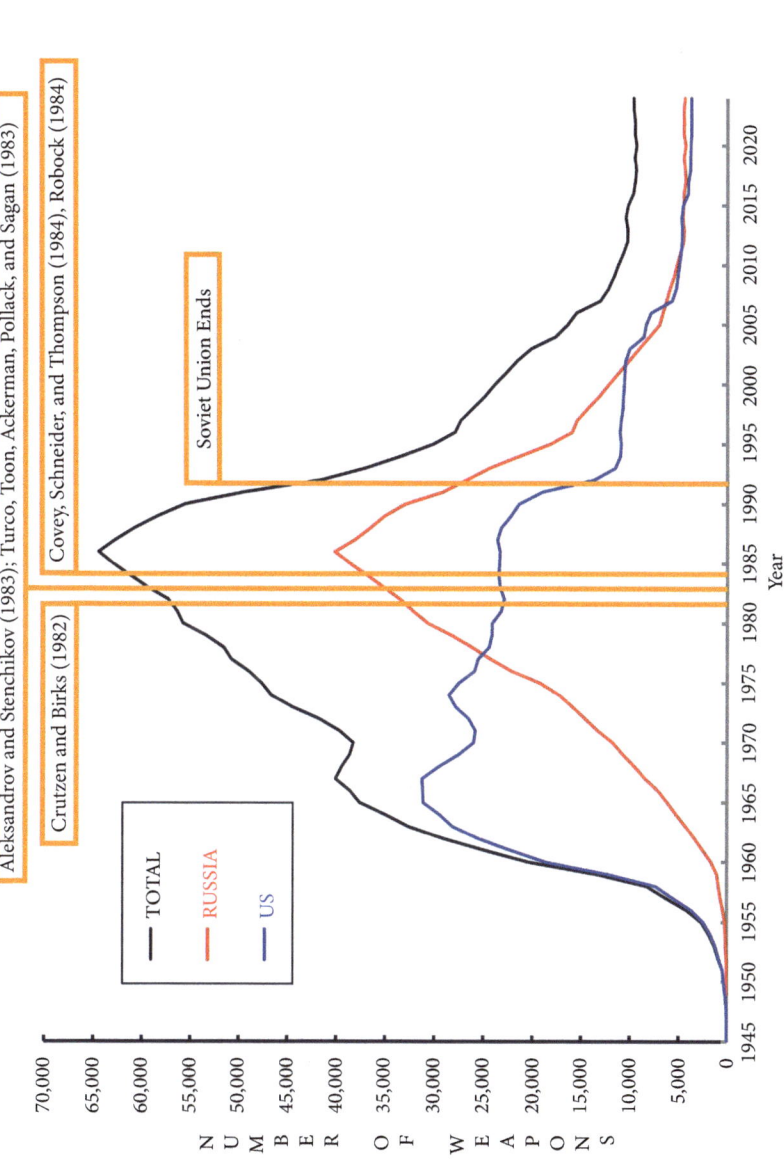

Figure 7.1 Time series of the total number of deployed nuclear weapons on Earth, which after about 2005 excludes large numbers of weapons in storage waiting to be dismantled. Total includes all nine nuclear-weapon states. See Figure 7.2 for the seven nuclear states not shown here, which have many fewer weapons.

Data from https://fas.org/initiative/status-world-nuclear-forces/ as of April 22, 2024 father was also in the Navy, working in the United States to interpret those photos to estimate steel production from Japanese factories. He went to Japan right after the war ended, arriving in Nagasaki and driving to different cities to get ground truth for the Navy estimates. Brian's father-in-law was a graduate student at the University of California at Berkeley who participated in research leading to the first atomic weapons. All three men were certain the bombing helped end the war with the minimum of fatalities, a widespread belief at the end of the war. However, many analysts today believe the bombings were unnecessary and that the military losses and the entry of the Soviet Union into the war, independent of the conventional raids, would have driven the war to its end just as rapidly.

Luis Alvarez composed a poignant letter to his young son Walter while flying on the chase plane after the bomb was dropped on Hiroshima on August 6, 1945 (National Archives):

I have been in this B-29 for eight hours so far, and we won't be back for another five or six. . . .

The story of our mission will probably be well known to everyone by the time you read this, but at the moment only the crews of our three B-29s, and the unfortunate residents of the Hiroshima district in Japan are aware of what has happened to aerial warfare. . . . A single plane disguised as a friendly transport can now wipe out a city. That means to me that nations will have to get along together in a friendly fashion, or suffer the consequences of sudden sneak attacks which can cripple them overnight.

What regrets I have about being a party to killing and maiming thousands of Japanese civilians this morning are tempered with the hope that this terrible weapon we have created may bring the countries of the world [together] and prevent further wars. Alfred Nobel thought that his invention of high explosives would have that effect, by making wars too terrible, but unfortunately it had just the opposite reaction. Our new destructive force is so many thousands of times worse that it may realize Nobel's dream.

Following the end of World War II, a long buildup of nuclear weapons occurred, as illustrated in Figure 7.1, as did a proliferation of new nuclear states, as illustrated in Figures 7.2 and 7.3. By the early 1960s, people such as Albert Schweitzer, a Nobel Peace Prize winner, and Linus Pauling, a Nobel Prize winner in Chemistry and in Peace, led massive protests of nuclear weapons testing in the atmosphere due to buildup of long-lived radioactive materials worldwide. For example, strontium-90 was detected in milk and in children's bones and teeth.

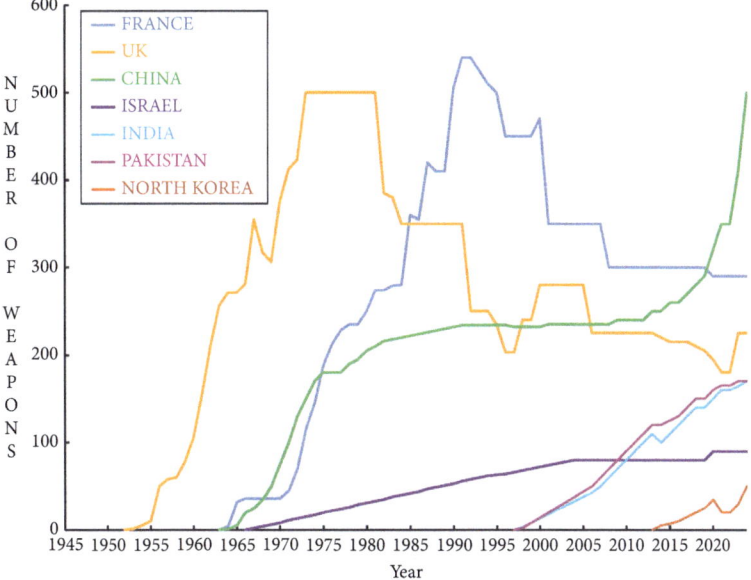

Figure 7.2 Time series of nuclear weapons from all nuclear-weapons states except the United States and Russia, which are shown in Figure 7.1.

Data from https://fas.org/initiative/status-world-nuclear-forces/ as of April 22, 2024

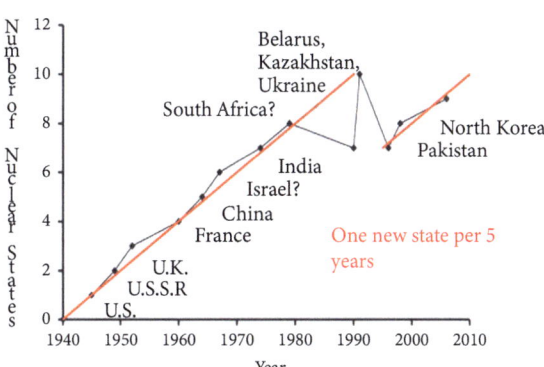

Figure 7.3 The dates when various nations obtained a nuclear warhead, mainly based on when they first tested a weapon, are shown. For Israel and South Africa, the evidence for tests is controversial, so an estimate for when they had a usable weapon is given. Belarus, Kazakhstan, and Ukraine inherited weapons from the Soviet Union and transferred them to Russia in the 1990s. South Africa gave up its weapons in the 1990s. The red lines represent one new nuclear state every 5 years. Toon et al. (2017).

In a September 25, 1961, address to the UN General Assembly, President John F. Kennedy, said:

> *Today, every inhabitant of this planet must contemplate the day when this planet may no longer be habitable. Every man, woman and child lives under a nuclear sword of Damocles, hanging by the slenderest of threads, capable of being cut at any moment by accident or miscalculation or by madness. The weapons of war must be abolished before they abolish us.*

The Cuban Missile Crisis of 1962 further convinced the public and political leaders that nuclear weapons were out of control. These protests about nuclear contamination and the Cuban Missile Crisis led to the 1963 Treaty Banning Nuclear Weapon Tests in the Atmosphere, in Outer Space and Under Water. Within 6 months, 100 countries had signed the treaty, but non-signatories China and France continued to conduct atmospheric tests, with China's last test in 1980.

President Kennedy worried in a speech to the American people in 1963 on the Nuclear Test Ban Treaty that

> *A war today or tomorrow, if it led to nuclear war, would not be like any war in history. A full-scale nuclear exchange, lasting less than 60 minutes, with the weapons now in existence, could wipe out more than 300 million Americans, Europeans, and Russians, as well as untold numbers elsewhere. And the survivors, as Chairman Khrushchev warned the Communist Chinese, "the survivors would envy the dead." For they would inherit a world so devastated by explosions and poison and fire that today we cannot even conceive of its horrors. So let us try to turn the world away from war. Let us make the most of this opportunity, and every opportunity, to reduce tension, to slow down the perilous nuclear arms race, and to check the world's slide toward final annihilation.*
>
> *I ask you to stop and think for a moment what it would mean to have nuclear weapons in so many hands, in the hands of countries large and small, stable and unstable, responsible and irresponsible, scattered throughout the world. There would be no rest for anyone then, no stability, no real security, and no chance of effective disarmament. There would only be the increased chance of accidental war, and an increased necessity for the great powers to involve themselves in what otherwise would be local conflicts.*

Unfortunately, Figure 7.1 shows little impact of the 1963 Test Ban Treaty on the total number of weapons, primarily because the Soviet Union built its arsenal following the treaty, and Figure 7.2 shows the Test Ban Treaty did not stop pro-liferation of nuclear weapons. The United States did, however, stop increasing the number of its weapons, which had reached an absurd level of 30,000. Even today there are only about 200 cities in Russia with more than 100,000 citizens. By 1960, the United States had enough weapons to attack these Russian cities with about 155 weapons used on each city. While many weapons might be used on military targets outside cities, one weapon each would suffice to destroy most of these cities, and many cities lie near military bases and would be destroyed as collateral damage.

Countries that have had nuclear weapon development plans, some extending into the 1990s, include Argentina, Australia, Brazil, Canada, Egypt, Iran, Iraq, Libya, Romania, South Korea, Spain, Sweden, Switzerland, Taiwan, and Yugoslavia (Albright et al. 1997). As Figure 7.3 indicates, South Africa did build nuclear weapons, but then gave them up, making it the only country to have ever discarded indigenous nuclear weapons. Belarus, Kazakhstan, and Ukraine inherited nuclear weapons from the disintegration of the Soviet Union but then gave them up. While the cascade Kennedy feared has not developed, steady proliferation continues today, with North Korea building and testing sophisticated weapons and several Middle Eastern countries threatening to build them.

Figure 7.1 shows that nuclear weapons peaked in 1986. In the early 1980s, there were worldwide demonstrations against nuclear weapons. At that time, there were 40,000 Soviet warheads, enough to attack each of the current 300 US cities with populations above 100,000 with 133 nuclear weapons. The world's citizens, and countries without nuclear weapons, recognized that having so many nuclear weapons was dangerous and unnecessary, even if politicians and bomb builders did not. President Ronald Reagan also inflamed opposition to the weapons, partly by proposing a ballistic missile shield, suggesting to some that a war could be won. The Soviet Union had many missiles aimed at targets in Europe. American missiles were deployed to Europe, where they would be able to hit the Soviet Union with almost no warning time. The threat to deploy missiles close to the Soviet border led to negotiations and to the downward trend in nuclear weapons after 1987. The Intermediate Range Nuclear Forces Treaty (INF) of 1987 required the United States and the Soviet Union to eliminate all their ground-launched ballistic and cruise missiles with ranges of 500 to 5,500 km (310 to 3,400 miles). This reduction was meant to stop the threat of an attack on Europe or the Soviet Union with little warning time. Every American president and every Russian leader since then has reduced the number of weapons. In the United States, Republican presidents, particularly the two George Bushes, made the largest reductions.

While President Reagan's image was closely associated with nuclear weapons, in fact he hated them and had a lifelong goal of getting rid of them. He and President Gorbachev tried to achieve this goal. While they did not succeed, they put the world on the path to reduce the numbers of weapons up until recently.

In 1985, before the INF treaty, Ronald Reagan gave an interview to the *New York Times*, noting

> *A great many reputable scientists are telling us, that such a war could just end up in no victory for anyone because we would wipe out the earth as we know it. And if you think back to . . . natural calamities - back in the last century, in the 1800's, . . . volcanoes— we saw the weather so changed that there was snow in July in many temperate countries. And they called it the year in which there was no summer. Now if one volcano can do that, what are we talking about with the whole nuclear exchange, the nuclear winter that scientists have been talking about? It's possible.*[1]

Soviet leader Mikhail Gorbachev, reflecting on the INF Treaty, observed in 2000:

> *Models made by Russian and American scientists showed that a nuclear war would result in nuclear winter that would be extremely destructive to all life on Earth; the knowledge of that was a great stimulus to us, to people of honor and morality, to act.* (Hertsgaard 2000)

The research they refer to is the central theme of this book. In the context of today's world, where science and fact are challenged by politicians, it is noteworthy that the work of the science community influenced the thinking of Presidents Reagan and Gorbachev.

Figures 7.1 and 7.2 end with about 10,000 total strategic nuclear weapons remaining partitioned, as shown in Figure 7.4. The reserve weapons in the Russian arsenal may include approximately 2,000 tactical weapons. There are several thousand weapons in storage waiting to be dismantled.

There are efforts to eliminate nuclear weapons. The United Nations Treaty on the Prohibition of Nuclear Weapons, developed in the UN General Assembly where it could not be vetoed by the states with nuclear weapons, was ratified on January 22, 2021. Unfortunately, none of the states with nuclear weapons and none of the states under their protection have signed the treaty[2]. There are numerous countries "protected" by the nuclear powers, including the non-nuclear European countries in NATO as well as Australia, Japan, and South Korea. Russia also has defense treaties with several countries. The United Nations Treaty on the Prohibition of Nuclear Weapons prohibitions are outlined in Table 7.1.

---

[1] *New York Times* interview, February 2, 1985, https://www.nytimes.com/1985/02/12/world/transcript-of-interview-with-president-on-a-range-of-issues.html?pagewanted=all
[2] United Nations Office for Disarmament Affairs, Treaty on the prohibition of nuclear weapons, https://www.un.org/disarmament/wmd/nuclear/tpnw/

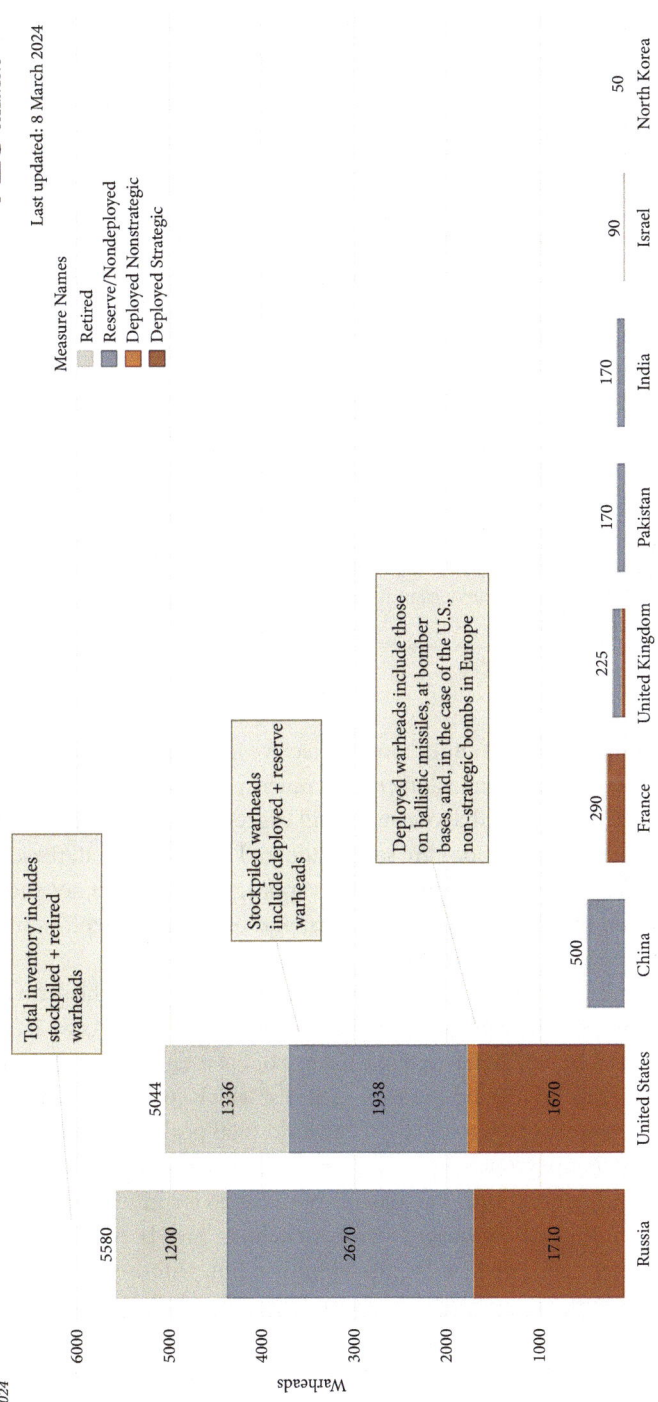

Figure 7.4 Estimated global nuclear weapons inventories, 2024, from Federation of American Scientists

Used by permission. https://fas.org/initiative/status-world-nuclear-forces/

Table 7.1 The First Article of the Treaty on the Prohibition of Nuclear Weapons

Article 1

Prohibitions

1. Each State Party undertakes never under any circumstances to:
   - (a) Develop, test, produce, manufacture, otherwise acquire, possess or stockpile nuclear weapons or other nuclear explosive devices;
   - (b) Transfer to any recipient whatsoever nuclear weapons or other nuclear explosive devices or control over such weapons or explosive devices directly or indirectly;
   - (c) Receive the transfer of or control over nuclear weapons or other nuclear explosive devices directly or indirectly;
   - (d) Use or threaten to use nuclear weapons or other nuclear explosive devices;
   - (e) Assist, encourage or induce, in any way, anyone to engage in any activity prohibited to a State Party under this Treaty;
   - (f) Seek or receive any assistance, in any way, from anyone to engage in any activity prohibited to a State Party under this Treaty;
   - (g) Allow any stationing, installation or deployment of any nuclear weapons or other nuclear explosive devices in its territory or at any place under its jurisdiction or control.

Three of the most common questions we receive about nuclear weapons are: Will one nuclear weapon explosion create a nuclear winter? Is there a threshold number of explosions such that one more will trigger a nuclear winter? *and* Can you tell me how many weapons are safe to allow? The answer to all these questions is "no." There is no safe number of weapons, the effects do not suddenly switch on when one more explosion occurs, and one nuclear explosion is not going to create a global catastrophe.

Still the number of nuclear weapons used in a conflict is one of the main things we need to know to evaluate the consequences of a nuclear war. As we have discussed, Figures 7.1 and 7.4 illustrate the history of the total number of warheads and the number in Russia and the United States, which currently control more than 91% of the weapons worldwide. The world total peaked at around 65,000 warheads in 1986, five years before the disintegration of the Soviet Union. While momentum and international agreements have led to a continuing downward trend, that trend is not continuing. Recent declines have been warheads that are being dismantled. The first Trump administration withdrew from the INF and from the Open Skies Treaty. The Open Skies Treaty allowed signatories to overfly other countries with unarmed aircraft with little notice to verify military activities. Unfortunately, many weapons remain on high alert, and rather than develop new treaties to reduce the dangers of nuclear weapons, both Russia and the United States have initiated expensive efforts to not only renew their nuclear arsenals but also to develop dangerous new types of weapons. For example, Russian President Vladimir Putin bragged in 2015 that Russia has developed a drone submarine (sometimes called a torpedo) that can carry nuclear weapons into American harbors. The yields of such weapons are not known, but based on the size of the torpedoes they could be a few megatons, or possibly much larger. The United States is developing, and Russia is using, hypersonic delivery systems to make it more difficult to intercept nuclear warheads and to reduce the warning times of an attack.

The 2024 world arsenal shown in Figure 7.1 is slightly less than 10,000 weapons, about 15% of the peak in 1986. Figure 7.4 illustrates the disposition of the strategic weapons. A significant number of strategic weapons, estimated to be about 1,800 in the United States and 1,800 in Russia, are being dismantled. The remaining approximately 10,000 are either deployed or in reserve in the United States, Russia, and other nuclear powers. Treaties restricting nuclear weapons only impact the United States and Russia. The Strategic Offensive Reductions Treaty (SORT) reached by Presidents George W. Bush and Vladimir Putin, and in place from 2003 to 2011, regulated the number of deployed strategic warheads to not exceed 1,700 to 2,200 for each country by December 31, 2012. That treaty introduced two new concepts, both undesirable. First, instead of regulating delivery systems, which are relatively easily observed from space, or from aircraft using the Open Skies Treaty, the treaty regulated warheads, which are difficult to monitor because they are so small. Second, rather than counting all warheads, SORT only considered those warheads that are deployed on strategic launch vehicles. Therefore, many warheads, about equal in number to those deployed, are in storage and might be mobilized in an emergency but are not counted. Additionally, nuclear weapons used for battlefield conflict, called tactical nuclear weapons, were not regulated by SORT. Such tactical weapons can have yields as high as strategic weapons. Russia may have about 1,600 tactical weapons, which could be used in a conflict involving Europe.

The New Strategic Arms Reduction Treaty (New START), developed under Presidents Barack Obama and Dmitry Medvedev and in place after 2011, resulted in similar numbers of deployed strategic warheads to those in the SORT treaty. Of the 10,000 total warheads believed to exist among all nuclear powers in 2020, only about 3,400 deployed strategic weapons were regulated under New START. Unfortunately, New START only regulates deployed launchers. The United States also has 20,000 plutonium pits in storage that could be converted to weapons. More than 1,300 unregulated weapons are owned by France, China, the United Kingdom, Israel, Pakistan, India, and North Korea together. Although the first Trump administration attempted to abandon the New START treaty, as soon as President Biden took office in January 2021, he reached an agreement with Russia to extend it for five years. However, Vladimir Putin suspended the New START treaty in 2023, in response the NATO countries supplying weapons to Ukraine during its war to repel a Russian invasion. The Trump administration also stopped reporting the numbers of US weapons in 2019, making it more difficult to track the number of weapons, but reporting was restarted in 2020, but then suspended again.

7.2 Who Has Nuclear Weapons, and How Could They Be Delivered?

The yields of the weapons possessed by the United States and Russia are listed in Table 7.2. These yields have been simplified from data based on specific warheads compiled by Kristensen et al. (2024a, b). The 800 warheads on the US intercontinental ballistic missiles (ICBMs) have yields of 300 kt (200 warheads) to 335 kt (600 warheads), with an average yield of 326 kt. Of these 800 warheads, 400 are deployed on the 400 US missile launchers. There are an additional 50 ICBMs in storage that could be deployed to empty silos. There are 1920 warheads available for submarine-launched missiles (SLBMs). About 1511 of these have a yield of 90 kt, and 384 have a yield of 455 kt, with an average yield of 162 kt. About 970 of the submarine-based warheads are deployed on 14 US submarines. The 14 submarines could carry as many as 280 missiles, but only 240 are currently deployed. The US submarines now carry about 25 tactical weapons of yield about 8 kt. These tactical weapons are a dangerous new escalation. The 2018 Nuclear Posture Review suggests these weapons may possibly be used to provide a prompt response option in cases of regional conventional aggression; however, some conjecture their goal is to provide the opportunity to use tactical nuclear weapons to break the taboo on nuclear weapon use. Some also suggested they were needed because Russia has more tactical weapons than the United States.

The United States also has strategic weapons for about sixty-six nuclear-capable bombers. In the New START treaty, each deployed bomber, whether it is nuclear or not, is counted as one nuclear weapon, though in fact many deployed bombers are not nuclear capable and those that are can carry many nuclear bombs or nuclear cruise missiles. The yields on these bomber weapons are variable. It is estimated that 528 nuclear warheads are carried by air-launched cruise missiles, whose yields are variable from 5 to 150 kt. Another 322 gravity bombs have variable yields from 5 kt up to 1200 kt. About 300 strategic nuclear weapons are deployed at bomber bases, and 488 are stored. Another 100 tactical nuclear weapons with yields up to 170 kt are on NATO fighter-bomber bases outside the

Table 7.2 Warheads and Yields for Deployed, Stored, and Being Dismantled Weapons (Kristensen et al. 2024a, b). ICBMs = intercontinental ballistic missiles, SLBMs = submarine-launched ballistic missiles.

Delivery System/Number	Number of Warheads	Number Deployed	Number Stored	Number to Be Dismantled	Average Yield in kt (Range)
US ICBMs/400	800	400	400		326 (300–335)
US SLBMs/14,280	1920	970	950		162 (90–445)
US bomber/66	788	300	488		(<5–1200)
US total strategic	*3,508*	*1,670*	*1,838*	*1,336*	*Total deployed yield ~287 Mt; average yield ~210 kt*
US tactical	200	100	100		(<170)
Russia ICBMs/326	1,244	872	372		337 (100–800)
Russia SLBMs/12,192	992	640	352		100 (100)
Russia bombers/67	586	198	388		Unknown
Russia total strategic	*2,822*	*1,710*	*1,112*	*1,200*	*Total deployed yield ~358 Mt; average yield ~236 kt*
Russia tactical	1,558				Unknown

United States, and another 100 tactical weapons are stored in the United States. In total, the United States has about 1770 deployed nuclear weapons, including tactical weapons, and 1938 in storage, for a total stockpile of 3708 weapons, including tactical weapons. The total yield of deployed weapons on ICBMs and submarines is about 287 Mt with an average yield of 210 kt.

The estimated yields of Russian nuclear weapons are listed in Table 7.2. These have been simplified from the data of Kristensen et al. (2024b), which assign yields to specific weapons. The yields of Russian warheads are not as well known as those of the United States because Russia does not publish the details of its strategic forces. Russia has an estimated 1244 warheads on ICBMs, with about 872 warheads deployed and the remainder in storage. The ICBM warheads are carried on 326 missiles. There are more warheads than missiles because some missiles have multiple independently targeted re-entry vehicles (MIRV). It is thought that 816 warheads have yields of about 100 kt, 78 have yields of about 800 kt, and 340 have yields of about 500 or 800 kt. The average yield of the warheads on ICBMs is about 337 kt, assuming the upper limit when a range of yields is given. Russia has about 992 warheads meant for SLBMs, with yields of 100 kt. About 640 SLBMs are deployed. Russia has 12 operational submarines that can carry 192 missiles. Russia also has about 586 warheads with unknown yields for about 67 operational bombers. About 200 of these warheads are deployed.

Figure 7.3 outlines the history of nuclear proliferation. Up until the mid-1980s, about one new nuclear state appeared every 5 years. Following the start of the build-down of nuclear weapons by the United States and Russia in 1986, several states abandoned their arsenals or stopped nuclear weapons programs that were under consideration or development. Unfortunately, proliferation was renewed in 1998 when India and then Pakistan tested nuclear weapons. As Figure 7.3 indicates, the world now may be back on the trend of increasing new nuclear states. Figure 7.4 illustrates the arsenals of the countries with nuclear weapons other than the United States and Russia (Kristensen et al. 2024c). Table 7.3 provides information on the arsenals of these countries (Kristensen et al. 2024c).

It is very difficult to determine the numbers or yields of weapons in most of these countries. Britain and France had been slowly reducing their arsenals, but Britain announced in 2021 that it planned to increase its arsenal. China is building new submarines and constructing new launch sites for ground-based missiles. The Israeli arsenal and its trends are not well known. Pakistan and India have been increasing their arsenals. It is projected that India and Pakistan will have 200 nuclear warheads each by about 2025, placing them in the same range of warheads as Britain and France. The yields of the weapons in most of these programs are not known. Britain, France, and China have weapons with yields above 100 kt. However, on the basis of their nuclear tests, India and Pakistan have weapons with yields similar to those of the US weapons used in World War II, around 10–40 kt, or even less. However, both Pakistani and Indian analysts claim they have higher-yield weapons.

Parts of the nuclear story are encouraging. Much of the world is ridding itself of nuclear weapons. Figure 7.5 shows the history of the development of nuclear weapons free zones (NWFZs), and Figure 7.6 illustrates the regions of the world in nuclear free zones. Almost one-third of the human population now lives in regions in which the United Nations recognizes treaties banning nuclear weapons.

The Treaty for the Prohibition of Nuclear Weapons in Latin America and the Caribbean, also known as the Treaty of Tlatelolco, went into effect in 1968 and

Table 7.3 Estimated Numbers of Weapons and Their Yields in Nuclear States Other Than the United States and Russia (Kristensen et al. 2024c; 2023a, b, c; Kristensen and Korda 2022a, b, 2021). ICBMs = intercontinental ballistic missiles, SLBMs = submarine-launched ballistic missiles.

Country	Number of Warheads Deployed/ Inventory	Yield per Warhead kt	Deployed Yield/Total Yield Mt	Types of Delivery Systems
United Kingdom	120/225	100 kt	12/22.5	SLBMs
China	0/500	~6 @ 4,000–5,000 most @ 200–300	0/~150	Aircraft, ICBMs, SLBMs
France	280/290	50 @ < 300 kt 240 @ 100 kt	~40	Land and carrier aircraft, SLBMs
India	0/170	10–40 based on tests	?	Aircraft, ICBMs, SLBMs
Israel	0/90	?	?	Aircraft, land-based ballistic missiles, possibly sea-based cruise missiles
North Korea	0/50	140–250 kt tested	?	Land-based ballistic missiles, SLBMs
Pakistan	0/170	5–40 based on tests	?	Aircraft, land-based ballistic missiles, cruise missiles

was eventually signed by all 33 independent nations of Latin America and the Caribbean, including Cuba. However, Brazil and Argentina reserved the right to conduct "peaceful nuclear weapons explosions," and islands such as Puerto Rico, British Virgin Islands, and Guadeloupe associated with nuclear weapons states are excluded. Africa established a NWFZ under the Treaty of Pelindaba, which took effect in 2009. As a result of the additional treaties for Antarctica and the Rarotonga Treaty involving Australia, New Zealand, and several island nations in the Pacific, the entire Southern Hemisphere (with the exception of islands associated with states with nuclear weapons and international waters) is a NWFZ. The Association of Southeast Asian Nations, including Brunei, Darussalam, Cambodia, Indonesia, Laos, Malaysia, Myanmar, Philippines, Singapore, Thailand, and Vietnam, became a NWFZ in 1997 under the Bangkok Treaty. There is also a Central Asian nuclear free treaty signed by Kazakhstan, Kyrgyzstan, Tajikistan, Turkmenistan, and Uzbekistan, which entered into force in 2009. Mongolia declared itself a NWFZ in 1992, and it was formally recognized

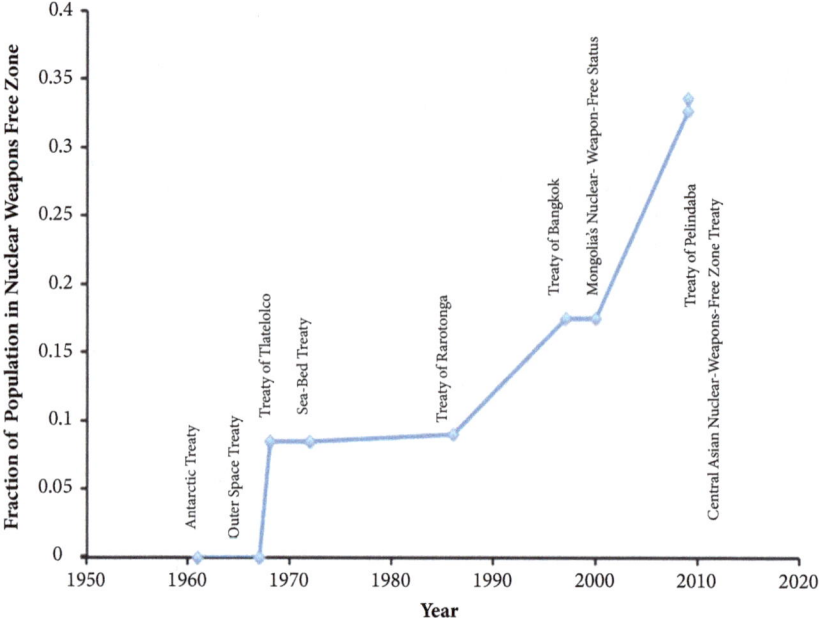

Figure 7.5 Treaties that create zones free of nuclear weapons. About one-third of the world population now lives in nuclear-free zones. From Toon et al. (2017).

as such in 2012 by the five original nuclear weapons states: the United States, China, France, the United Kingdom, and Russia.

Comparison of Figures 7.1 and 7.2 raises the question: how do countries know how many weapons are enough? None of the nuclear weapon states has stated its criteria for answering this question. In the case of the United States, analyses in the late 1940s identified militarily important industrial sites in the Soviet Union and concluded these could be destroyed by the delivery of 100 atomic weapons that were similar in yield to the weapons used in Hiroshima and Nagasaki (Eden 2004) and perhaps in the arsenals of present-day Pakistan, India, and North Korea. Of course, weapons may not explode or may never reach their target, and so it was estimated that 200 weapons would be needed for the United States to destroy the Soviet Union. Despite this estimate, the US arsenal rose to more than 150 times this many weapons, and with typically more than 10 times the average yield per weapon. One reason for the bloated number of weapons may be the competition between Russia and the United States and attempts by each to dominate the other so that a first strike might overwhelm the adversary (Lieber and Press 2006). In the case of a first strike, each missile of the adversary must

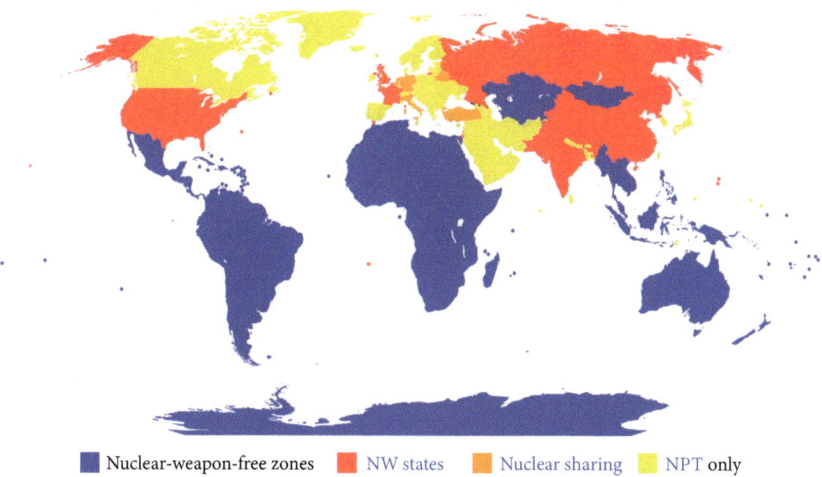

Nuclear-weapon-free zones   NW states   Nuclear sharing   NPT only

Figure 7.6 All the Southern Hemisphere except for a few small islands controlled by nuclear weapons states is a nuclear-free area. This is a map of the world with different kinds of nuclear proliferation and nuclear-weapon-free zones (NWFZs) as defined by United Nations Resolution 3472 from 1975. Only states as a whole are represented, not separate territorial parts of them, even if they would belong to a different group. Blue indicates NWFZs by international treaty, including territories that belong to a nuclear weapons state that has agreed the territory is subject to a zone. Red are nuclear weapons states and territories belonging to them that are not in any NWFZ. Orange indicates nuclear sharing (United States or Russian nuclear arsenal stationed there for host country use in wartime). Yellow is none of the above but party to the Non-Proliferation Treaty (NPT).

https://en.wikipedia.org/wiki/Nuclear-weapon-free_zone#/media/File:Nwfz.svg

be targeted with multiple warheads to be certain they are destroyed. The opponent must then obtain even more warheads to be able to attack each of the enemy's missiles with more than one warhead. This competition to outnumber the other side leads to exponential growth in warheads. Alternative suggestions for the large numbers of weapons include retention of old weapons when new ones are developed, concern over potential development of effective defenses, competition for funding to support the nuclear infrastructure driven by the military-industrial complex that President Eisenhower warned of in his farewell address, political posturing, lack of planning, or lack of thought about the size of the arsenal needed or concern that there would be a high failure rate of the weapons.

Chapter 8
Scenarios for War and Near Misses

8.1 Near Misses

"The Russians were very supportive of us after the revolution," said Fidel Castro to Alan in Havana, Cuba, on December 15, 2011 (Figure 8.1). "They gave us oil and food, and every ship they promised arrived, for many years. So, when they wanted to put missiles here, eventually I said 'yes,' and once I said 'yes,' I could not back down and had to support them. They needed the missiles to counteract the American ones in Turkey. I did not want them here but was convinced to agree to them."

Thus, Fidel Castro outlined the start of the Cuban Missile Crisis, when the world came close to nuclear war. The Cuban Missile Crisis of October 1962 is one of many times, that we know of (and there may be many others that are classified), that the existence of nuclear weapons and some combination of confusion, technical failures, carelessness, mistakes, and panic have brought the world close to a nuclear war. The excellent extensive compilation of these cases by Sarah Witmer (2017) is summarized in Table 8.1, listing only the most egregious ones.

There are several patterns to the near misses in Table 8.1. Most serious are cases where someone has mistakenly thought that a first strike had begun during a period of tension in the world. Often these have been cases where a warning system had been put into place recently and malfunctioned or was misinterpreted. For example, on November 5, 1956, the British and French were attacking Egypt over control of the Suez Canal when "U.S. defense forces receive word of what seems to be a Soviet invasion: unidentified aircraft are flying over Turkey, Soviet MIGs are flying over Syria, a British bomber has been shot down over Syria, and the Soviet fleet is moving through the Dardanelles in northwestern Turkey. The American military fears that this might trigger a NATO nuclear strike against the USSR. All four signs of invasion are later disproven by various unrelated events: the unidentified aircraft were actually a flight of swans, the MIGs were a routine air force escort for the Syrian president as he returned from Moscow, the British bomber was forced down for mechanical reasons, and the Soviet fleet was engaging in routine exercises" (Witmer 2017).

A more serious example occurred on November 9, 1974. "President Carter's National Security Adviser, Zbigniew Brzezinski, receives a phone call at 3 a.m.

Earth in Flames. Owen Brian Toon and Alan Robock, Oxford University Press. © Oxford University Press (2025).
DOI: 10.1093/9780197799734.003.0008

Figure 8.1 Left to right: Antonio Castro Soto del Valle, Fidel Castro Diaz-Balart (Fidel's sons), Fidel Castro Ruz, his wife, Dalia Soto del Valle, Juan Carlos Antuña (Cuban scientist, former student of Alan Robock), author Alan Robock (with gift of Rutgers baseball cap for the Comandante), Betty Muñoz (interpreter), Peter Agre (2003 Nobel Prize in Chemistry), and Tomás Gutiérrez (head of the Cuban Institute of Meteorology—their Weather Service) at a meeting on December 15, 2011

by his military assistant adviser General William Odom, announcing that 2200 Soviet missiles have been launched at the United States. President Carter has less than ten minutes to order retaliation. Just as Brzezinski is about to call the president, Odom calls again to say it was a false alarm: someone mistakenly placed military exercise tapes in the operational missile detection computer system" (Witmer 2017).

Several other incidents involving failures of the warning system occurred. On September 26, 1983, a radar system malfunctioned during a tense period of the Cold War and reported a launch of several missiles. Famously, the Soviet officer on duty, Lieutenant Colonel Stanislav Petrov, suspected an error and did not report the warning. He later received international awards in the West for preventing a nuclear war, and a docudrama about him featuring Kevin Costner called him "the man who saved the world." On January 25, 1995, "Russian early warning radar detects a scientific rocket launch off the coast of Norway (which the U.S. had informed Russia about beforehand), and mistakenly identifies it as an American submarine-launched ballistic missile. Russian nuclear forces jump to full alert, with President Boris Yeltsin retrieving the nuclear launch codes and preparing for a retaliatory launch. Russian satellites monitoring American missile fields prove that the missile is not headed for Russia, and a strike is called off" (Witmer 2017).

Another class of mistakes includes accidents in which aircraft crash or malfunction (releasing nuclear weapons), missile silos experience fires, submarines collide, or nuclear weapons are loaded onto aircraft by mistake. Of course, close calls on nuclear wars are not limited to the United States and Russia. India and Pakistan have had conflicts in which nuclear war was threatened in May–June 1999 and December 2001–October 2002. American Secretary of State Mike

Table 8.1 Serious incidents when the world came close to nuclear war. Listed are incidents with a severity scale of 3–5, where 3 is a specific and serious risk, possibly leading to escalation with the other state, that requires complex resolution; 4 is a serious risk to wider public that has potential to cause widespread casualties and damage beyond military personnel and property or to cause escalation in conflict; and 5 is nuclear devices detonate and cause casualties or confrontation nearly leads to the use of nuclear devices.

November 10, 1950: Plane accidentally drops nuclear weapon
American plane in Canada, Severity: 4
> A B-36B bomber experiencing mechanical failure drops its Mark 4 atomic bomb over Quebec. Its conventional explosives detonate when it lands in a river, scattering nearly 45 kg (100 pounds) of uranium.

March 10, 1956: Plane carrying nuclear weapons disappears
United States, Severity: 3
> A B-47 carrying two types of nuclear capsules from Florida to a base overseas loses contact over the Mediterranean, and is never found.

July 27, 1956: Plane crashes into bomb storage
American base in United Kingdom, Severity: 3
> A B-47 bomber skids off the runway on landing and rips into a storage igloo containing Mark 6 atomic bombs before exploding. The bombs do not detonate.

November 5, 1956: False alarm of Soviet attack (flight of swans) during Suez Crisis
United States, Soviet Union, United Kingdom, and France, Severity: 3

January 31, 1958: Plane fire on runway turns bomb into a 3600 kg (8,000 pound) block of radioactive metal.
American base in Morocco, Severity: 3

February 5, 1958: Plane collision drops nuclear weapon over water and the bomb is never found
United States, Severity: 3

March 11, 1958: B-47 bomber accidentally drops a Mark 6 atomic bomb into backyard in South Carolina
United States, Severity: 4

November 4, 1958: B-47 bomber crash into a field near Abilene, Texas with Mark 39 hydrogen bombs on board
United States, Severity: 3

October 15, 1959: B-52 with nuclear weapons on board collides over Hardinsberg, Kentucky with refueling aircraft
United States, Severity: 4

October 5, 1960: False alarm (radar misinterprets moonrise over Norway as missiles) suggests attack
American base in Greenland, Severity: 3

January 19, 1961: B-52 bomber crash over Monticello, Utah with nuclear weapons on board
United States, Severity: 3

January 24, 1961: Plane crash drops two nuclear bombs into Goldsboro, North Carolina
United States, Severity: 4

November 24, 1961: Communications failure (actually lines down in Colorado) suggests enemy attack
United States, Severity: 3

August 23, 1962: Navigational error of nuclear B-52 into Soviet airspace
United States, Severity: 3

October 1962
Throughout the Cuban Missile Crisis of October 1962, miscommunications due to the chaotic nature of the issue at hand as well as sheer carelessness led to multiple near-nuclear confrontations.
Severity: 5

Throughout the month: Miscommunication possibly signals attack
United States and European allies, Severity: 3

Throughout the month: Prolonged UK exercise possibly signals attack
United Kingdom, Severity: 3

October 24, 1962: Satellite explosion misinterpreted as attack
Soviet Union, Severity: 3

October 25, 1962: alarm of sabotage (actually a bear entering a Wisconsin base) almost leads to attack
United States, Severity: 4

continued

Table 8.1 *continued*

October 26, 1962: missile test possibly signals attack to Soviets
United States, Severity: 3

October 26, 1962: Unannounced missile test causes false alarm of attack to Americans
United States, Severity: 3

October 26, 1962: Nuclear missile left alone with launch codes
United States, Severity: 3
> A single operator could have singlehandedly launched a nuclear-armed missile.

October 27, 1962 is now commonly referred to as "Black Saturday" as it was the most dangerous day of the Cuban Missile Crisis, when both the United States and the Soviet Union came close to initiating nuclear attack multiple times.

Cruise missiles pointed at the United States
Soviet base in Cuba, Severity: 4
> In the early morning of October 27, the Soviets deploy nuclear cruise missiles in firing position to within 24 km (15 miles) of the US naval base at Guantanamo Bay. The US remains completely unaware.

Wartime radio frequencies signal war
Soviet Union, Severity: 4

U2 spy plane enters Soviet air space
United States, Severity: 4
> Meanwhile, an American U2 spy plane enters Soviet air space, attracting the attention of Soviet MIG interceptors, which are ordered to shoot the plane down. American fighter planes loaded with nuclear missiles and ordered to shoot at their own discretion are sent to escort the U2 plane back to American ground.

U2 spy plane shot down over Cuba
United States, Severity: 5

Submarine almost launches nuclear torpedo
Soviet Union, Severity: 5
> Perhaps most seriously, eleven US Navy destroyers and aircraft carrier U.S.S. Randolph corner a nuclear-armed Soviet submarine near Cuba. Authorized to launch nuclear torpedoes without express permission from Moscow, two of the three submarine officers onboard vote to launch. The third officer, Vasili Arkhipov, refuses to authorize the launch. Had any other officer been in Arkhipov's place—whether one who agreed with the two other officers, or one who was more easily pressured by the other officers to authorize the launch—nuclear war likely would have occurred.

October 28, 1962: Simulation tape interpreted as attack from Cuba
United States, Severity: 4

October 28, 1962: False alarm and miscommunication suggest missile attack
United States, Severity: 3

November 2, 1962: Captured secret agent gives false alarm of nuclear attack
United Kingdom, Severity: 4

December 5, 1965: Plane falls off the deck of the U.S.S. Ticonderoga aircraft carrier
American plane over the Pacific Ocean, Severity: 3
> A bomber carrying a nuclear weapon, rolls into the ocean. Pilot, plane, and weapon are never found.

January 17, 1966: B-52 bomber collides with a plane refueling it mid-air spewing radioactive material
American plane in Spain, Severity: 4
> Seven crew members are killed in the crash, and American military crew brought in to clean up after the crash show high rates of radiation-related illnesses today. Spanish people from the area also contracted cancer and other illnesses at higher rates, and sections of Palomares remain highly radioactive today.

May 23, 1967: Communications failure suggests nuclear attack
United States, Severity: 3
> Multiple early warning radar sites around the world go offline, leading the US to fear that the Soviets have disabled American radar in the first stage of a nuclear attack. Nuclear bombers prepare to take flight until it is determined that a solar flare knocked out the radar systems.

January 21, 1968: American B-52 bomber crash spews radioactive material
American base in Greenland, Severity: 4
> The plane crashes into sea ice, causing the conventional explosives in all four bombs to detonate and radioactive material to be spewed into the ocean. Had the plane hit Thule Air Base, American Strategic Air Command would likely have assumed attack and retaliated.

April 11, 1968: Nuclear submarine sinks
Soviet submarine in Pacific Ocean, Severity: 3

November 15, 1969: American and Soviet submarines collide
Barents Sea, Severity: 4

October 24-25, 1973: False alarm signals nuclear attack during Arab-Israeli War
United States, Severity: 3

August 1, 1974: Unfit president holds power to launch nuclear attack
United States, Severity: 3
> In his last weeks in office during the Watergate crisis, Nixon is depressed, drinking heavily, and extremely unstable.

November 9, 1979: False alarm nearly leads to nuclear strike
United States, Severity: 4
> Someone mistakenly placed military exercise tapes in the operational missile detection computer system.

continued

Table 8.1 *continued*

June 3 & 6, 1980: Faulty computer chip announces missile attack
United States, Severity: 3

September 18, 1980: Fire at a Titan II ICBM nuclear missile silo in Damascus, Arkansas
United States, Severity: 4
    A missile repairman drops a socket from a wrench. The fuel explodes, killing an airman, and catapults the warhead out of the silo.

September 26, 1983: Radar malfunction warns of missile attack
Soviet Union, Severity: 3
    The Soviet officer on duty, Stanislav Petrov, suspects a malfunction and does not call for a retaliatory Soviet strike.

November 2-11, 1983: NATO military exercise Able Archer 83 interpreted as attack preparations.
Soviet Union, Severity: 4

January 10, 1984: Malfunction causes nuclear-armed missile to almost launch
United States, Severity: 3

August 19-21, 1991: Coup leaders confiscate nuclear briefcases from Gorbachev
Soviet Union, Severity: 4

January 25, 1995: Scientific rocket launch from Norway interpreted as nuclear armed missile
Russia, Severity: 4

May-June, 1999: Conflict almost includes nuclear weapons
India and Pakistan, Severity: 5
    The incident escalates until both sides threaten to use nuclear weapons. The crisis is defused by mediation by President Clinton.

December 2001-October 2002: Conflict almost includes nuclear weapons
Pakistan, Severity: 3

August 29-30, 2007: Nuclear missiles accidentally loaded onto plane
United States, Severity: 3

May 23, 2008: Fire in missile silo burns unnoticed
United States, Severity: 4

July 28, 2012: Activists break into top-secret uranium warhead component production plant
United States, Severity: 3

August 5, 2014: Nuclear power plant sabotaged
Belgium, Severity: 4

Adapted from *Nuclear Close Calls* by Witmer (2017)

Pompeo stated that in 2019 he was involved in diffusing a situation in which both India and Pakistan thought the other was preparing for a nuclear attack (Pompeo 2023).

Another important type of close call has been due to leaders who appear unstable. In the failed coup in 1991 against Gorbachev, he lost control of the Soviet nuclear launch codes. During the resignation of US President Richard Nixon and at the end of US President Donald Trump's first term, there was widespread concern that they should not have sole control of their country's nuclear weapons.

The closest the world came to its second nuclear war was during the 1962 Cuban Missile Crisis. The Soviet Union had secretly installed 36 ballistic missiles in Cuba aimed at the United States, each with a 1 Mt warhead (67 times more powerful than the bomb dropped on Hiroshima). When the United States discovered them, President Kennedy demanded the Soviets remove them from Cuba. When there was not an immediate response, President Kennedy ordered a naval blockade of Cuba to stop any additional nuclear weapons from arriving. The resolution of the crisis involved several political close calls.

Daniel Ellsberg's job was to help design nuclear targeting strategies in the 1960s during the Kennedy administration. In his 2017 book, *The Doomsday Machine: Confessions of a Nuclear War Planner*, he explains how primitive the security measures were on the large number of deployed nuclear weapons, remaining so for decades after they were put in the field on submarines, on airplanes, and on missiles. When asked what he thought of the Stanley Kubrick movie *Dr. Strangelove*, released in 1964, which portrayed a nuclear war started by a madman combined with a secret Russian doomsday machine, he said he thought it was a documentary.

As detailed in Table 8.1, a US U-2 spy plane was shot down over Cuba by a Soviet anti-aircraft missile, which was the "only deliberate, acknowledged killing of an American soldier by Soviet troops in the entire Cold War" (Witmer 2017). The local Cuban military oversaw anti-aircraft weapons and, under orders from Castro, tried to shoot down lower-altitude US reconnaissance planes, which could have triggered a wider conflict, despite orders from the Soviet leader, Nikita Khrushchev, not to make any provocations.

Several Soviet submarines were located near Cuba and carried nuclear-armed torpedoes. US Navy ships decided it would be useful to force the submarines to the surface by throwing practice depth charges, similar to hand grenades, on them. They informed the Soviets that their submarines should surface, and they would be warned by the practice depth charges. Unfortunately, the Soviets were unable to contact their submarines with this information. Of course, the submarines feared they were being depth charged. Eventually, three of the

submarines did surface without incident, for which they were severely repri-
manded. However, one of these Soviet submarines almost launched nuclear
torpedoes. The captain of the submarine and the political officer on board voted
to launch. Fortunately, unlike on the other Soviet submarines, there was another
officer on board, Vasili Arkhipov, with the same rank as the captain. He vetoed
the use of the nuclear weapons. Arkhipov's wife was later proud that her husband
had become known as "the man who saved the world."

At various points during the crisis, the United States threatened the Soviets with an armed invasion of Cuba, not knowing there were over 100 tactical
nuclear weapons, warheads included, in Cuba in case of such an event. There
were also far more Soviet troops in Cuba than were known by the United States
at the time. Such an invasion would almost certainly have triggered a nuclear
war.

Neither Kennedy nor the Soviet leader Nikita Khrushchev wanted a nuclear
war, and they were able to make a secret agreement that the United States would
soon remove their nuclear missiles from Turkey if the Soviets would first remove
their nuclear missiles from Cuba and end the crisis.

8.2 Launch on Warning

The reason that mistakes could lead to a nuclear war is that there is very little time to verify that a mistake has been made. A nuclear conflict may begin
due to a failure of the launch-on-warning protocols used by Russia and the
United States. In the United States, a single individual, the US president, can
order a launch of a large fraction of the nuclear arsenal without approval by
anyone else, and in a moment's notice, as described by many authors (e.g., Perry
and Collina 2019). To protect the land-based intercontinental ballistic missiles
(ICBMs) from loss, they are ready to be launched quickly. The US president
is constantly accompanied by a military officer carrying the launch codes and a
device to activate the launch, which is called the football even though it looks like
a conventional briefcase. The US thought they detected a Russian or Soviet first
strike in progress at least three times, and the Russians believed they detected
an American first strike in progress at least twice, as outlined in Table 8.1. It is
only through good luck that in each case the evidence that the attack was real
was not convincing enough to launch a counterattack.

Unfortunately, the system is still vulnerable to mistaken detections of an
attack. In the past, many false alarms occurred when new warning systems were
put in place and malfunctioned. To get around US anti-ballistic missile defenses,
Russia is currently developing new types of weapons, such as nuclear cruise
missiles that can be in continuous flight and attack from unexpected directions,

drone submarines that can carry very large nuclear weapons into America harbors, and hypersonic gliders that can attack with little warning. The evolution of weapons systems, which could be prevented by new treaties, will require new warning systems that can react more quickly, leading to even less time for a human or a machine to decide whether the attack is real. It is even possible that we will soon turn over decisions to launch the weapons to artificial intelligence. In that case, software errors may lead to nuclear war. We are sure the reader has experienced many malfunctions with their laptops and cell phones, even though these systems have been tested by millions of users over decades of development. New systems introduce new opportunities for malfunctions. Even current detection systems are inadequate. For instance, Russia relies almost entirely on radar to detect a first strike. But radars are restricted to line-of-sight detection and therefore have limited range. The United States augments radars with satellite observations of infrared signatures of missiles, but Russia has limited capabilities for satellite detection and therefore less time to decide if an attack is real.

8.3 Scenarios for Wars

Currently, nine nations have nuclear weapons: the United States, Russia, China, the United Kingdom, France, India, Pakistan, Israel, and North Korea. Five are completely in Asia (Toon et al. 2017), two in Europe, and one in North America. Russia is in both Europe and Asia. There are no nuclear weapons based in the Southern Hemisphere. There are 186 nations in the world without nuclear weapons. Here we discuss how those countries with weapons might come to use them in warfare. Some of the current nuclear nations are allies, and nuclear war between them is not likely. Although the United States and United Kingdom have fought two bloody wars in the past, the US Revolutionary War and the War of 1812, they worked together to create the first atomic bomb and through the North Atlantic Treaty Organization (NATO) are allies, so we do not consider a scenario of a US-UK nuclear war credible. Similarly, the United Kingdom and France have a long history of warfare, but they are now allies. The United States and Israel are allies. Although, more in flux recently than the alliances just mentioned, it does not seem to us that China and Russia, China and Pakistan, or China and North Korea are likely to have a nuclear war since they are economically and politically entwined. But there are places and relationships in the world that are more volatile and could succumb to a nuclear war. We consider several of the most dangerous cases here, but many similar scenarios are possible, and the ones presented are simply plausible examples of scenarios.

There are various ideas in the scenarios described below that come from the Cuban Missile Crisis and the near misses in Table 8.1. The most likely cause for a nuclear war is that during a period of heightened tensions a mistake of some sort will be misread as an attack, and as a result a counterattack will be initiated and carried out.

8.4.1 India and Pakistan

India and Pakistan share a 3,323 km (2,065 mile) border, of which 962 km (598 miles) are in Jammu and Kashmir (Figure 8.2). Ongoing skirmishes along the Jammu-Kashmiri border, terrorist raids (such as the one from Pakistan into Mumbai in 2008), or even the US incursion into Pakistan in 2011 to capture and kill Osama bin Laden (which could have been mistaken as an Indian attack) could all lead to nuclear war. India and Pakistan have had four conventional

Figure 8.2 NASA astronaut photograph ISS045-E-27,869, acquired on September 23, 2015. "The port city of Karachi is the bright cluster of lights facing the Arabian Sea, which appears completely black. City lights and the dark color of dense agriculture closely track with the great curves of the Indus valley. For scale, the distance from Karachi to the foothills of the Himalaya Mountains is 1,160 kilometers (720 miles). This photograph shows one of the few places on Earth where an international boundary can be seen at night. The winding border between Pakistan and India is lit by security lights that have a distinct orange tone."
https://earthobservatory.nasa.gov/IOTD/view.php?id=86725

wars (in 1947, 1965, 1971, and 1999) since their partition in 1947. India claims that there were 3,289 ceasefire violations by the Pakistan Army in 2019, the most since the 2003 India-Pakistan border truce. In February 2019, an Indian fighter jet was shot down over Pakistan, but the pilot was soon returned to India. Nevertheless, the continued unrest in this region makes it one of the most likely places in the world for a combination of unfortunate circumstances to lead to nuclear war.

As we discussed in detail (Toon et al. 2019), there are many credible scenarios in which conflicts between India and Pakistan could escalate to nuclear war. This would require substantial provocation. India has a policy of no first use of nuclear weapons, except in response to an attack with biological or chemical weapons. Pakistan has similarly declared that it would not use nuclear weapons except to stop an invasion by conventional means or if it were attacked by nuclear weapons. We, however, have laid out how this might happen in 2025 when we expect India and Pakistan to each have 200–250 nuclear weapons.

Here we outline one possible scenario. It begins with a terrorist attack on the Indian parliament, as happened on December 13, 2001. In this case, though, there would be massive fatalities among members of India's government. (This is not meant to suggest that other scenarios are any less likely that could begin with India starting the conflict, but the possibility of any such scenario is the issue.) As happened in January 2002, we assume that India and Pakistan mobilize their troops within a few weeks of the terrorist attack. As did happen in 2002, Indian Army units deploy along the border. As might happen in a future conflict, the Indian Army crosses the Line of Control (the de facto border) in Kashmir. The next day Pakistani conventional forces respond but are overwhelmed. Pakistan then uses tactical nuclear weapons inside its own border and in Kashmir against Indian tanks. This could be so shocking that fighting stops and further war is averted. But in the "fog of war," a term first used by Hale (1896), panic, poor communications, and rigid procedural rules continue the escalation. As Pakistan continues to use tactical nuclear weapons, India attacks 20 military garrisons and airfields collocated with urban areas using nuclear weapons. The next day Pakistan continues to use tactical weapons and uses ballistic and cruise missiles on garrisons in 20 Indian cities and 10 Indian naval bases and airfields in urban areas. On the same day, India attacks 10 Pakistani Navy, Army, and Air Force bases in urban areas. Things are now out of control, and in total about one-half of the total Pakistani and Indian arsenals are used on cities. An additional 20 Pakistani and 25 Indian weapons are used on military targets in rural areas. Overall, Pakistan uses its entire nuclear arsenal, although a fraction of the weapons would fail. A fraction of India's weapons would not work either. India would reserve 100 weapons as a deterrence against China (ignoring the fact that the 150 weapons they just used against Pakistan did not work to deter them). Overall, depending on whether India and Pakistan used 15 kt bombs as they tested in 1998 (the size that devastated Hiroshima in 1945) or have been able to develop 50 kt or 100 kt weapons, this scenario could produce 50 to 125 million fatalities from the direct effects of the explosions. Nuclear-ignited fires could release 16 to 37 Tg of black carbon in smoke, which would threaten the rest of the world with fatalities that could range into the billions due to starvation, as is discussed later.

8.4.2 United States and North Korea

A possible scenario would begin when North Korea tests a long-range missile and the United States and South Korea shoot the missile down with an anti-ballistic missile defense system. North Korea responds by launching another conventionally armed missile at the Terminal High Altitude Area Defense (THAAD) missile base in South Korea, where anti-ballistic missile defense systems are located. Then the United States launches non-nuclear Tomahawk cruise missiles at North Korean nuclear missile bases, and general conventional war breaks out. North Korea barrages Seoul with conventional artillery, and the US and South Korean forces invade North Korea and threaten to defeat the regime within weeks.

The North Koreans reply with their complete nuclear arsenal, 30 weapons with yields of 100–200 kt against US bases in South Korea, Japan, and Guam, some of which are in cities, and another 20 weapons on US cities on the West Coast. The United States replies with 80 submarine-borne nuclear weapons of 100 kt or 475 kt against North Korea. Of these 80 weapons, 50 are used against counterforce targets (North Korean military), 20 are used against war-supporting industries, and 10 are used against command and control.

A United States–North Korea scenario involves the risk of entraining Russia or China into the conflict. What if China or Russia detects missiles flying over their territory on the way to North Korea and thinks it is a ruse to attack them? What if weapons go off course and land in China or Russia? What if plumes of radioactivity blow into Vladivostok, close to the North Korean border, or into Chinese cities? What if China decides to help defend their North Korean ally? To our knowledge, no studies exist of the consequences of this type of war for casualties due to the direct impact of the weapons, nor of the possible effects on the rest of the world.

In a terrifying page-turner, *Nuclear War: A Scenario*, Annie Jacobsen (2024) describes another nuclear war scenario involving North Korea, and its aftermath, which includes a nuclear winter. As we show later, the climate impacts depend on how much smoke ends up in the upper atmosphere and not how it gets there.

8.4.3 United States and China

This hypothetical war begins when the nationalist Taiwanese government declares its permanent resistance to integration with mainland China. This crosses a red line for China, which responds with conventional military action, such as a blockade. The United States assists Taiwan, including deployment of US aircraft carriers in the Taiwan straits, which presents a direct threat to China. China responds by mobilizing its conventional and anti-ship ballistic missiles and deploying troops where it might attack Taiwan.

Many possible interactions could lead to further escalation. North Korea might take this as an opportunity to attack South Korea with conventional forces. Or a US ship patrolling in the Taiwan Strait might encounter a Chinese Navy ship on a closing course, and a panicking officer might launch weapons at the ship. Or the United States might shoot down a Chinese surveillance aircraft or sink Chinese anti-ship submarines. China might respond by sinking a US carrier or a smaller vessel in the US fleet. Or China could attack critical satellites on which the US Navy depends for survival of its fleet in the region. Except for its nuclear arsenal, the United States may lack the military capabilities to promptly eliminate China's anti-satellite threat. The United States might not be able to attack the anti-satellite base with stealth aircraft due to China's dense anti-air defense systems. In these scenarios, the only US tool to protect its fleet from annihilation might be low-yield submarine-launched nuclear missiles fired from the Western Pacific.

The conflict could then escalate to large-scale nuclear war. The United States might launch submarine nuclear forces (100 kt or 475 kt per warhead, with 5% failure rate) against 450 Chinese counterforce targets, including command and control, war-supporting industry, and military bases, some of which are in urban areas. China retaliates against US cities with 48 submarine warheads on 4 submarines (200–300 kt each). Other potential Chinese targets could include 20 US military bases in South Korea, Japan, and Guam.

8.4.4 Baltic Conflict Leads United States, France, United Kingdom, and Russia into All-Out War

Unfortunately, one of the most probable locations for a nuclear war is in the countries bordering Russia. Unlike Ukraine, the Baltic countries are NATO members, and NATO is bound to defend them. Like Ukraine, the Baltic countries were once part of the Soviet Union and still have Russian-speaking citizens among their populations. The Russians could send disguised forces or even special operations forces into the Baltic states claiming to support the Russian population, as they have done in Ukraine, Georgia, and other former Soviet republics. These incursions would result in conflict between the Russian forces and local military, police, and civilians. The conflict could escalate to US and NATO involvement, with the invocation of NATO Article 5, which provides that if a NATO ally is attacked all other NATO members must come to its defense. Lethal battles with conventional weapons between Russian and US/NATO troops would ensue. The conflict could widen to engage large-scale military formations (ground, air, naval in the European theater), including full-scale alert of all nuclear forces on both sides.

As a Russian defeat by the superior NATO conventional forces looms, Russian forces would retreat toward Russian territory. Russia might then employ a single tactical nuclear weapon (50 kt on a Polish military facility) to warn NATO not to cross into Russian territory, as a demonstration of resolve according to their "escalate to de-escalate" dogma. Escalation to tactical nuclear exchange in the European theater on retreating Russian forces might then occur involving 100 US/NATO B-61 bombs (0.3–170 kt each; likely high-yield, low-altitude airbursts) and 250–500 Russian tactical missiles and aircraft weapons (5–200 kt each; Russia has approximately 600 tactical nuclear warheads of unknown yield on bombers and missiles; Kristensen et al. 2024b) against major NATO military facilities throughout Europe. Escalation to strategic nuclear war between United States/United Kingdom/France and Russia could quickly follow.

8.4.5 Hacking

A hacker launches one missile or one set of US land-based missiles (1–50 weapons, 300 kt or 335 kt) or one to several Russian land-based missiles (100 kt to 800 kt). This could trigger a launch on warning, with all land-based missiles on both sides launched.

Hacking could occur in several ways. In one case, with the help of an insider or embedded malware, unauthorized third parties could hack and fire strategic missiles on launch-ready alert. Is it possible to hack a squadron of 50 weapons or the entire 400-weapon US (or 150-silo Russian) inventory? This might trigger an early warning response. In a second case, would it be possible to hack launch codes from 10 to 15 high-level command posts? The use of these codes would potentially compromise the nation's entire nuclear arsenal.

Hacking does not have to be in the United States or Russia. For example, there could be hacking or intrusion in Pakistan after tactical weapons are delegated to unit commanders. Or there could be hacking and delegation with unauthorized action at multiple levels in multiple nuclear states. Fortunately, in some countries such as India and Pakistan, the nuclear weapons may not be mounted on missiles or even assembled, so hacking would be difficult.

8.4.6 Other Possibilities

The Russians claim to have nuclear-powered, long-range torpedoes (called "Poseidon" in Russia). Rumors suggest very high yields and release of radioactive debris over target areas. One could imagine another Russian attempt at nuclear blackmail, in which the Russian leader announces such weapons are in the harbors of New York, Washington, Los Angeles, San Francisco, and Miami (or Boston or Baltimore or Seattle) and will be detonated unless the United States agrees to their demands. Would this be seen as a first strike and trigger an all-out nuclear war?

The large number of deployed nuclear weapons discussed in Chapter 7 are not just sitting around waiting for someone to decide what to do. Due to launch-on-warning protocols requiring the ability to quickly launch the arsenals, there are detailed war plans and predetermined targets. There may be a nuclear weapon aimed at you right now, as discussed in the next chapter.

Sidebar 8.1 Alan's Visit with Fidel Castro

Fidel Castro was the leader of Cuba when Alan had the surreal experience of a meeting with him, during which Fidel explained his version of what happened during the Cuban Missile Crisis. Alan's work on nuclear winter had been discovered by Fidel Castro in 2010, and through his connection with former PhD student Juan Carlos Antuña, Alan was invited to Havana to give a talk about the work, with the Comandante in the audience. Fidel had the talk recorded and shown on nationwide television in prime time the next day, on September 15, 2010. That day, they gave Alan a private, guided tour of Old Havana. As they went into one of the many nice, restored hotels downtown, Alan asked to see a TV, as his entire 90-minute presentation was being broadcast nationwide and he wanted to see it. But the TV in the hotel bar was showing a Julia Roberts film, so Alan asked them to change the channel. Sure enough, there he was, with Fidel watching his presentation. This reinforced Alan's understanding that if we really want to get the word out on the climatic consequences of nuclear weapons, it has to be in a dramatic film with big stars like Julia Roberts to get people to pay attention. It must touch people's emotions, not just their intellect. We are writing this book in hopes that someone will make these issues and our work more accessible to the public.

Later, on September 22, 2010, Fidel blogged about Alan's visit: "The nuclear winter theory, developed and brought to its current stage by the eminent researcher and professor from Rutgers University, New Jersey, Dr. Alan Robock (a modest scientist who prefers to recognize the merits of his colleagues rather than his own), has proven its veracity. . . . We promised the professor that we would spread the information he had provided us about the nuclear winter theory—a topic we know a little about due to our concern over the possible outbreak of a global nuclear war, a concern that drew us to his lecture—in a language that even eight-year-old Cuban children could understand."

A year later, Alan traveled to Cuba in a scientific exchange visit organized by the American Association for the Advancement of Science. Alan told Fidel's son, Fidel Castro Diaz-Balart, a PhD scientist whom he had met on the previous visit, the scientific advisor to the Cuban government, and known as Fidelito, that the last time he met him Alan did not bring any gifts, as he did not know he would be meeting Fidelito or his father. Now Alan had two Rutgers baseball caps, one for him and one for his father. "Which one is for my father?" Fidelito asked. Alan told him to just pick one he liked and give the other to his father. He said, "He likes blue, and you can just give it to him." In shock, Alan asked how, and Fidelito said we are going to meet him after lunch. Maybe Fidelito just liked the black one (Figure 8.3). In the meeting, Alan sat across a table from Fidel Castro for more than 3 hours while Fidel told his life story, including what he thought about the Cuban Missile Crisis (Figure 8.1). See the blue Rutgers hat in Alan's hand as he prepared to present it to Fidel. Alan proposed to him a meeting in Havana the next year on the fiftieth anniversary of the Cuban Missile Crisis with the foreign ministers of the nine nuclear nations, to include educating them about nuclear winter and the climatic consequences of even a nuclear war between two of the smaller nuclear states. Fidel agreed, but unfortunately never carried through with organizing it.

Alan's encounter with Fidel Castro suggests that politicians and policy advisors involved in controlling nuclear weapons can learn from the past and change their ideas.

Figure 8.3 Alan Robock and Fidel Castro Diaz-Balart and his Rutgers baseball cap after the meeting in Figure 8.1.

Chapter 9
Are You Being Targeted with a Nuclear Weapon?

About one-third of the world's population lives in nuclear-free zones, as shown in Figures 7.5 and 7.6. They are not likely to be targeted with nuclear weapons. However, slightly more than half of the world's population lives in areas that could be attacked by nuclear weapons. Almost 1 billion people, about 12.5% of the world's population, live in NATO countries. The United States has military bases or deployed troops in about 80 countries. The largest number of American troops outside NATO are in South Korea and Japan, but there are also significant bases or facilities in Bahrain, Kuwait, Iraq, and Qatar. About 200 million people, 2.5% of world population, live in the Collective Security Treaty Organization (Armenia, Belarus, Kazakhstan, Kyrgyzstan, Russia, and Tajikistan), but there are also Russian enclaves in Crimea, Azerbaijan, Georgia, and Uzbekistan. India's population of 1.4 billion and Pakistan's 220 million constitute about 20% of the world's population, while China's 1.4 billion is also close to 20%. If you live in any of these countries near a military base, a government nuclear lab, a large commercial airport, a large oil refinery, a major power plant, a busy harbor or train yard, a military contractor, or a government center, it is possible that a nuclear weapon is either aimed at you right now or that there are plans to target you.

Of course, in an attack the number of weapons used, and their actual targets depend on how the war is fought, which cannot be known in advance. This uncertainty produces a wide range of possible targets and scenarios for war. Here we discuss the criteria that may be used to identify targets and provide some examples of targets. No country has divulged their actual target lists, and even the criteria for choosing targets is uncertain.

9.1 Rational Wars. Can Targets Be Limited in Scope?

Scenarios for how a war is fought vary according to the self-imposed rules of the conflict, which are highly classified secrets. The countries involved in the conflict may not even have the same rules. During WWII, the rules evolved during the conflict with little regard to the "legal" rules of war as nations sought to achieve victory.

Earth in Flames. Owen Brian Toon and Alan Robock, Oxford University Press. © Oxford University Press (2025). DOI: 10.1093/9780197799734.003.0009

Some analysts argue that there could be "rational" wars. In such a conflict, a small number of nuclear weapons, possibly just one, would be used to demonstrate their power, and then negotiations would quickly bring the war to an end in favor of the country using the weapons. During the first Trump administration, low-yield nuclear weapons were put on submarines, possibly in the belief that such weapons could be used to end a war in a limited conflict (probably involving a country that did not have nuclear weapons of their own). At the time of writing this book, Russia seems to be considering using tactical nuclear weapons during its war in Ukraine. While of little practical value against dispersed military targets, a few tactical nuclear weapon explosions might scare Ukrainians into surrender before seeing their cities obliterated. Fortunately, so far Russia seems to be taking no actions to put its nuclear weapons on an alert status or to be readying them for use, possibly because they fear world condemnation for using a nuclear weapon.

Some war games have explored the idea of a "rational war." War games are conducted between analysts pretending to represent different sides in a conflict or between representatives of different countries to explore how they might make decisions in a war. Basically, a scenario for a war is proposed, and then the people involved in the game pretend to be generals and politicians conducting the war. This is something like playing a video game. A war game exercise in Sri Lanka during 2013 organized by the US Naval Postgraduate School involving retired senior military and civilian analysts from India and Pakistan found that "a limited war in South Asia will escalate rapidly into a full war with a high potential for nuclear exchange" (Khan and French 2013). Currently, Pakistan and India depend on the United States and other countries to intercede if a dispute threatens to get out of control. Pakistan is a much smaller country than India and in danger of being overrun in a conventional attack. If India tried to invade Pakistan or if Pakistan thought India were about to invade, Pakistan might feel it had to use its nuclear weapons before they were overrun. In short, a limited war for India could be seen in Pakistan as a full-scale war. During the war game of 2013, Pakistani participants quickly started to ready nuclear missile launchers and release warheads to their Strategic Forces Commands. It was suggested that Pakistan might use tactical nuclear weapons inside its own territory not only to stop an invading army but also to keep the war from expanding. However, Indian participants responded that the use of a single tactical nuclear weapon even inside Pakistan would result in a massive retaliation. The plausibility of this scenario occurring is one reason that nuclear conflict between India and Pakistan is of such concern.

An important fallacy in thoughts of rational wars is that there will be time to carefully consider the results of your actions. A nuclear war would likely be different from all conventional wars, not only in the degree of damage but also in duration. During World War II, the many years of conflict allowed for development of new weapons and changes in strategy. The United States initially avoided large-scale bombing of urban areas and attempted to use precision bombing with the Norden bombsight because indiscriminate bombing had been banned by the League of Nations and President Roosevelt had cited it as a problem before the United States entered the war. However, the bombsight was found to have a circular error probable (CEP, meaning 50% of the bombs would land within the CEP) of about 360 m (1,200 feet), making the bombsite useless for attacking industrial and military targets in urban areas with high accuracy. The Army Air Force estimated 90% of their bombs failed to destroy their targets. Also, bombing during the daytime so that the bombsite could be used made the aircraft easy to attack. During the last years of the conflict, the United States and the United Kingdom attacked cities at night with hundreds of aircraft using incendiary bombs. The goal was to burn down large urban areas, particularly in Japan, but also in European cities such as Dresden and Hamburg. The Tokyo fire-bombing of March 9 and 10, 1945, is thought to have killed more people than died at Hiroshima or Nagasaki.

The argument for bombing civilians in WWII was that it would force the enemy to surrender. But there is no evidence that it worked. In the current Russia-Ukraine war, Russians are bombing civilian targets possibly for the same reason, and so far, as we write this, these tactics are only making Ukrainians more resolved to keep fighting and generating international support for Ukraine.

In contrast, a nuclear war between nuclear armed adversaries is likely to last for a period of hours or days. Even if an attempt were made to slowly escalate war starting with tactical weapons, the fear of an all-out attack is likely to quickly drive the conflict to full use of all weapons. Military planners face a dilemma. If a city in their country were destroyed in a nuclear attack, would military planners wait for politicians to discuss what to do in response? Or would military planners assume they have only one chance to destroy the enemy and attack all important targets using all their weapons before the enemy launches more weapons at them? These issues likely have already been resolved in American and Russian launch plans that have been prepared for the leaders of these countries, since they currently have only a few minutes to decide to launch many of their weapons if they are warned that they are already under attack (Jacobsen 2024).

9.2 Will the Laws of War Protect City Dwellers from Being Targeted?

A possibly important issue is whether targeting nuclear weapons on urban areas is legal. This is only possibly important because there is little evidence that during war combatants care about rules of war. Articles 51, 52, and 53 of Protocol 1 of the Geneva Convention of 1977 ban attacks on civilian population centers. They also ban putting military targets in civilian areas for the purpose of protecting them from attack. Article 56 bans attacks on dams, dikes, and nuclear electrical generating stations unless they are being used for military purposes, because the damage might extend well away from the target. Because of these counterbalancing rules, it may be legal to attack urban areas by claiming there are military facilities located there. The United States has not ratified its initial recognition of Protocol 1. Russia withdrew from Protocol 1 in 2019. Therefore, it is not clear if attacks on urban areas would be avoided by either the United States or Russia, even assuming they were trying to conduct legal warfare. The war in Ukraine obviously involves many Russian attacks on civilian targets such as railroad stations and apartment buildings. In the case of railroad stations, Russia claims they are being used for military transport and are legitimate targets.

In an extensive review of the evolution of the laws of war, Richard (2016), then deputy staff judge advocate at the United States Strategic Command, noted his opinion:

> *The atomic strikes on Hiroshima and Nagasaki were lawful under the laws of war existing in 1945. These attacks also represent the only two instances of atomic weapon strikes in history. They illustrate the definitional confusion existing with respect to the U.S. classification of lawful military objects exclusive of civilian objects. If some historians believe President Truman was engaged in "self-deception" in labeling Hiroshima a military target, those historians may be equally perplexed by modern American usage of law-of-war target labels.*

Richard also observes that in his opinion

> *Under the mid-twentieth century military's target lexicon. . . . the President's [President Truman] understanding of Hiroshima as a military target was accurate.*
>
> *Although Hiroshima was not a military base as understood today, it was a "military city" as it housed the 2d Army Headquarters, which commanded the defense of all of southern Japan. The city was also a communications center, a storage point, and an assembly area for troops.*

According to Richard (2016), US nuclear war-fighting plans up through the 1950s, that have been released, mimic the ideas from WWII with use of nuclear weapons on the majority of Soviet cities. They also indicate that nuclear weapons would not be held in reserve, precluding "no first strike" declarations. By the early 1960s, the concept of mutually assured destruction (MAD) arose. According to studies quoted by Richard (2016), massive retaliation for any attack included destruction of one-half to two-thirds of Soviet industrial capacity and one-quarter to one-third of the Soviet population. The basic concept behind MAD is that treaties and laws cannot be relied on to prevent a nuclear attack. Instead, fear of extreme damage to both military and civilian entities will prevent nuclear war. A corollary to this view is that nuclear weapons have no military value and are essentially useless militarily except for providing deterrence.

During the 1960s, many treaties, nuclear-free zones, Red Cross pronouncements, and UN resolutions were made trying to restrict the purposeful use of nuclear weapons on civilian populations. However, these rules generally recognize that civilian casualties are unavoidable when military and civilian targets are collocated. According to studies quoted by Richard (2016), the 1980 Presidential Directive 59 in the Carter administration did not explicitly target population, but despite this "all 200 of the largest Soviet cities and 80% of cities with populations over 25,000 contained targets for potential nuclear strikes." Apparently, these targeting lists were continued through the Reagan and Bush administrations.

According to Richard (2016), in the late 1980s targeting selections began to require oversite from judge advocates to ensure the targets were consistent with the rules of war. Recent debates and the 2015 Department of Defense Law of War Manual support the idea that attacks are possible on objects that are important for war fighting as well as war sustaining. For instance, targets that are important for funding war are legitimate if they can be directly connected to the military as opposed to general financial health of the society. In the context of the war in Ukraine, some Russians have threatened to use drone submarines to explode multi-megaton weapons upwind of the United Kingdom to spread radiation over a wide area. According to sources quoted by Richard (2016), the Russian drone submarines have the aim of preventing "military, economic or any other activities . . . for a long period of time" in the region attacked.

The law of war requires that attacks be limited to military targets. However, economic and industrial facilities are legitimate targets. While cities, by themselves, legally cannot be listed as targets, their economic and industrial facilities can be, and they may be targets in reprisal, which is a central theme of deterrence through massive retaliation and assured destruction. However, given the likely short duration of a nuclear war, it may not be possible to know the adversary's targets before the reprisal targets are selected and attacked. Richard (2016) noted:

> Ambiguity in the law of war can serve to improve deterrence by keeping adversaries uncertain as to the exact nature of potential responses to aggression.

9.3 Counterforce and Countervalue Wars

Two types of targets in nuclear wars are referred to as "countervalue" and "counterforce," in the sanitized language nuclear strategists use to insulate themselves from the horror they contemplate. In a countervalue war, the targets are chosen based on their intrinsic value to the targeted country. Such targets would include population centers, historical sites, and economic targets. In a counterforce war, targets are chosen based on their military value. Such targets would include military bases, but also oil refineries, power plants, government laboratories, war-supporting industries, airports, harbors, and government centers. It was found in the 1980s, when there were many high-yield weapons each of which could destroy a large area, that counterforce wars and countervalue wars did not differ greatly in their impacts on people because many of the counterforce targets are located in or near countervalue targets such as cities. At the current time, there are fewer weapons, and they have lower yields. Due to the lower yields of the weapons, the area damaged in an explosion is smaller. Therefore, countervalue and counterforce targets might not be located close enough in space to be attacked with one weapon, limiting incidental damage to urban areas. Unfortunately, we are not aware of any modern evaluations of the differences in casualties and other damage between countervalue and counterforce wars. Since many weapons have adjustable yields, the overlap between countervalue and counterforce wars is difficult to predict in advance of a war in which the yields are chosen. Some assume it is US policy not to attack any Russian civilian population centers even if they contain military targets. In this case, countervalue and counterforce wars would differ greatly.

9.4 Nuclear Targets in a First Strike

During the Cold War, both the United Stated and the Soviet Union feared that the other might gain nuclear superiority and initiate a first strike. The goal of a first strike is to destroy the enemy's military to such a degree that they could not respond and attack you back. Clearly, the primary targets in this case are the nuclear warheads of the enemy.

Whether a first strike was ever considered by responsible people on either side is not known. However, by the time of the Intermediate Range Nuclear Forces Treaty in 1987 Presidents Ronald Reagan and Michael Gorbachev agreed that "a nuclear war cannot be won and must never be fought." At present it is not possible to be certain you could stop an enemy from responding to a first strike for at least two reasons. First, some types of delivery systems cannot be located now with certainty. For example, submarines are difficult to locate, and both Russia and the United States likely believe that their submarines are not detectable and would survive a first strike. Likewise, once aircraft were airborne, they could be difficult to destroy before they launched their cruise missiles, which might also be difficult to destroy. Russia uses many mobile launchers, and it would be difficult to locate all of these if they scattered. Second, it would be a gamble to assume your missiles could destroy the enemy's missiles before the enemy could launch theirs. Nevertheless, the United States has never stated that it would not be first to use nuclear weapons, and Russia has changed its policy to allow nuclear first use if a conventional war should imperil the future of the state.

First-strike nuclear wars have obvious targets, mainly the missile silos, submarine bases, and military airfields where the opponent's nuclear weapons are located. The numbers of land-based intercontinental ballistic missiles (ICBMs) are given in Table 7.2 for Russia and the United States. Submarine-launched ballistic missiles (SLBMs) carrying nuclear weapons make it impossible to conduct a successful first strike unless submarine detection greatly improves, a situation referred to as the "ocean becoming transparent." Currently, deployed SLBMs can wait to launch their missiles until they can confirm that a war is underway. If they become vulnerable, or believe they are vulnerable, then they will need to launch their missiles as soon as warning of an attack is received, which increases the danger of an accidental war due to a false warning.

The New START Treaty, currently still being followed by the United States and Russia, limits the number of deployed strategic warheads to 1,550, but it counts bombers as one warhead, so the actual number of deployed weapons is near 2000 for each country. The treaty also specifies that there can be no more than 700 deployed ICBMs, SLBMs, and bombers, and no more than 800 such delivery systems including those that are not deployed. With twice as many warheads on each side as delivery systems, it becomes possible for there to be a first strike using two warheads on each delivery system target. Of course, if submarines cannot be found, and aircraft or ICBMs can be launched before they are destroyed, a first strike using two warheads per target will fail.

The United States is more vulnerable to a first strike than Russia because its 400 ICBMS are located in 450 silos whose locations are well known. Russia has about 329 land-based ICBMs (three of which do not carry nuclear weapons). Many of these are mobile and would be more difficult to destroy than American siloed weapons because their locations are not fixed. In a first strike by either side, it is unlikely that one offensive weapon would be enough for assured destruction of one land-based missile on its launcher. Therefore, Russia would need to use at least 900 of its strategic weapons to attack the 450 US silos with two warheads each. Russia has about 870 ICBM warheads deployed, so it would need to use ICBM warheads from storage or deployed bombers and submarine weapons to complete a first strike. However, it would still be left with about 810 deployed strategic nuclear warheads after the first strike and about 2,000 warheads in storage to use on other US or NATO targets. The United States would need to use 652 of its strategic weapons to attack the Russian ICBMs with two deployed warheads each. So, the United States would need to use not only deployed ICBMs but also weapons on submarines and bombers or to add warheads from storage to its ICBMs. After a first strike, the United States would have about 1018 deployed warheads left, but it would have another 2000 warheads in storage left for other targets. Additional weapons would need to be aimed at bomber bases in each country and ports for submarines. There are only a few locations where nuclear weapons are stored for bombers, but with advanced notice the aircraft could be deployed to commercial or military airfields or other locations. There are also only a few nuclear submarine bases. Most targets would be ICBMs. Although each country has about 2000 warheads in storage and some delivery systems that could be mobilized or augmented, it is likely that any attempt to make these stored weapons useable in advance would be detected, leading to a political crisis and alerting the other side to a possible first strike. While numerically possible to hit all these targets with two warheads, there are many other targets that are conventional military forces, and either side would run out of weapons before it ran out of targets.

A few analyses have been made of the consequences of a first strike. As pointed out by Perry and Collina among others, it is not wise to locate targets of a first strike on your territory. For instance, about 450 silos and control centers are located in the states of Colorado, Nebraska, Wyoming, Montana, and North Dakota. Their precise locations are obvious to any user of the internet. These missiles and silos are referred to as "sponges" by military analysts, since if you are not planning to use them in a first strike their only value is taking nuclear explosions away from apparently more important coastal cities. It would be smarter, from a military strategy point of view, to eliminate the land-based ICBMs by negotiation so that there would be no clear first strike targets in the United States or Russia. This would also remove the temptation to launch them on warning of an attack, given that warnings may not be correct. Without the ICBMs, both countries would still be left with plenty of warheads to destroy the other country.

9.5 Targets in Response to a First Strike

The most likely case for a war involving the United States, NATO, and Russia is a war by mistake. We now discuss in detail some of the targets that might occur in this situation. We assume that either Russia or the United States mistakenly thinks that the other country has launched its missiles in a first strike. In response, the country that thinks it is under attack launches its missiles, which in the case of a mistaken identification of a first strike would lead the other country to launch its missiles. In reality, it might take a few days for all of the nuclear weapons to be used, but the frailties of command-and-control systems would put pressure to use or lose weapons. Submarine weapons would not need to be used quickly assuming that their locations are not known to the enemy, but one might never know in advance if they had been detected. Some strategic weapons might fail, but the failure rate for US and Russian ICBMs is expected to be low, and the ability of anti-ballistic missile defenses to destroy incoming missiles is likely extremely poor. Aircraft-borne nuclear weapons, mainly in the form of cruise missiles, might have a high failure rate due to either aircraft maintenance issues, inability for aircraft to takeoff before incoming missiles destroy them on the ground, or destruction of the relatively slow cruise missiles by air defense systems.

If you are responding to a first strike, it makes little sense to use two of your weapons on each of the empty silos left over after the enemy launched their first strike, but you might use one weapon on each launch site in case a few ICBMs were not launched. There are many other important targets that would need to be destroyed to "win" a war. Of course, there are also many military targets that are not nuclear.

The current target lists are highly classified. A list of US targets from 1956 has been released. At that time there were no ICBMs, so many of the targets were military airfields. Another large class were industries that supported the military. There were also targets listed as population, railroad yards and equipment, ports, radio and television, and fuels at refineries. These types of targets remain likely at the current time, but some are controversial or now are given other names. For example, populations might now be called government centers or war-supporting industry. In Table 9.1, we delineate some possible targets in the United States, NATO, Russia, and their allies in a counterforce war in response to a first strike. We assume that ICBM silos are targeted with just one weapon, so this scenario maximizes damage on other types of targets. In addition to knowing the location of the target, it is important to know if the target is hardened. Missile silos, command-and-control centers, nuclear weapons storage sites, and some other targets are likely either underground, in caves, or covered in thick concrete. Such hardened targets require ground bursts and highly accurate targeting to destroy. Most targets—such as cities, army bases, or airfields—do not require high accuracy to destroy and will have greater damage done by airbursts whose shock waves propagate farther than those of ground bursts. However, targeting planners might still use ground bursts to attack cities since the radioactive fallout might make the local area uninhabitable.

Table 9.1 Potential targets in a counterforce war involving the United States, NATO, Russia, and their allies, assuming that nuclear launch facilities are targeted with one weapon.[g] Colored circles are keys to Figures 9.1–9.4. Red circles are military base targets. Black circles are energy-related targets. Yellow circles are regions where urban areas are likely located.

Type of Target		U.S. and Allies[a]	NATO[b]	Russia and Allies[c]
Allied military bases	🔴	23		47
Navy units[d]	🔴	213	49	54
Air Force units	🔴	137	137	202
Army units	🔴	134	111	111
ICBM silos or launchers	🔴	450	0	329
EMP		3		
Oil refineries[e]	⚫	146	106	21
Oil storage locations	⚫	12		
Power plants[i]				
Fossil fuel	⚫	120	54	133
Nuclear	⚫	32	36	9
Hydroelectric	⚫	9	3	66
Commercial airports	🟡	10	142	100
Military-supporting industry	🟡	56		99
Commercial harbors on the ocean	🟡	60		

continued

Table 9.1 *continued*

Type of Target	U.S. and Allies[a]	NATO[b]	Russia and Allies[c]
Railyards	70		25
National laboratories	46		44[f]
Government centers	51		85
Total	**1,572**	**638**	**1,325**

[a]Column includes targets in United States and Canada and other allies, not including European NATO. The number of targets in Allies are Australia (2), Bahrain (1), Britain (Diego Garcia) (1), Guam (3), Japan (8), Qatar (1), South Korea (6), and UAE (1).
[b]Canada is listed under the United States. The other NATO countries are Albania, Belgium, Bulgaria, Croatia, Czech Republic, Denmark, Estonia, Finland, France, Germany, Greece, Hungary, Iceland, Italy, Latvia, Lithuania, Luxembourg, Montenegro, Netherlands, North Macedonia, Norway, Poland, Portugal, Romania, Slovakia, Slovenia, Spain, Sweden, Turkey, and the United Kingdom.
[c]Allies include (# of targets in parentheses) Armenia (3), Moldova (1), Abkhazia (3), Tajikistan (2) Syria (2), Belarus (5), Crimea+ Ukraine (30), Kyrgyzstan (1).
[d]Naval units often include aircraft bases, and marines or other army units.
[e]We limited targets to oil refineries producing 2,000 or more barrels of refined products from crude oil per day.
[f]We identify these as locations of closed cities, which are cities where foreigners are not allowed to enter.
[g]Values without footnoted sources in the text were compiled by the authors from searches on the web.
[h]Includes silos that are empty or being upgraded.
[i]We assume that in the US and Canada only power plants with output above 1500 MW are targeted.

Of course, there are many variations of Table 9.1 that could be devised. One might decide that some of the targets don't merit the use of nuclear weapons but could be destroyed by conventional weapons. Or one might not attack government centers so that someone remains in charge to surrender or agree to a ceasefire. Table 9.1 should be considered a plausible list used as an example; actual targets are classified. It is likely that both military targets and population centers containing military targets would be attacked in a full-scale conflict.

For both counterforce and countervalue wars, we need to consider alliances of the United States, NATO, and Russia beyond their simple territorial definitions. We include Canada, a NATO country along with the United States in Table 9.1, because it shares a common border and many common companies and facilities.

The NATO countries other than Canada and the United States could be attacked using either strategic or tactical nuclear weapons. Perhaps tactical nuclear weapons might be defined as nuclear weapons whose numbers and yields are not well known. There is no clear difference between strategic and tactical weapons except for the range of their delivery vehicles

(Kristensen and Korda 2019). They may have the same range of yields. Generally, strategic weapons are delivered by long-range missiles or bombers, while tactical nuclear weapons are delivered by aircraft or missiles with shorter ranges. Tactical weapons also include battlefield weapons such as nuclear depth charges and artillery shells, but we will not consider the possible targets for those weapons since the targets are not located at a fixed location. The United States tested an artillery shell with the yield of the weapon use on Hiroshima in WWII, but no longer has such weapons. Russia has many tactical nuclear weapons, possibly approaching 2000 (Table 7.2) (Kristensen et al. 2024b). The United States has about 100 tactical nuclear weapons with yields up to 170 kt located in Europe as bombs and another 100 such weapons stored in the United States.

The United States has about 625 military bases that are overseas (514) or in US territories (111). Many of these are small or temporary. Many larger bases are included in our list of NATO bases. We assume that Russia will feel threatened by about 23 of these bases located in eight countries that are not part of NATO, as listed in Table 9.1. These countries are Australia, Bahrain, Britain (Diego Garcia), Guam (a U.S. territory), Japan, Qatar, South Korea, and the United Arab Emirates.

Russia is part of the Commonwealth of Independent States, which includes military cooperation. The other members are Armenia, Belarus, Kazakhstan, Kyrgyzstan, and Tajikistan. Russia has mutual defense agreements with South Ossetia and Abkhazia, which most of the world recognizes as parts of the independent country of Georgia. Russia also has extensive military bases in Crimea, which most of the world recognizes as part of Ukraine. Russia has military bases, most with more than a thousand personnel, in Armenia, Belarus, Georgia (South Ossetia, Abkhazia), Kazakhstan, Kyrgyzstan, Moldova (Transnistria), Syria, Tajikistan, and Ukraine (Crimea).[1] We consider this wider group of countries to contain potential targets. Russia may be currently constructing military bases in six countries in Africa: the Central African Republic, Egypt, Eritrea, Madagascar, Mozambique, and Sudan.

In Table 9.1, we chose targets up to the limit of nuclear weapons in Table 7.2. We assumed Russia and the United States will target ICBM launchers with single weapons, because we envision a conflict that starts with the mistaken assumption that the other side has launched a first strike. As mentioned earlier, in an actual first strike the initiator would use two weapons on each ICBM to assure destruction. However, in retaliation it makes little sense to attack empty silos with two weapons, but one might use one weapon in case not all the ICBMs had been launched. It is likely that many other targets would be attacked with two

---

[1] List of Russian Military Bases Abroad, https://en.wikipedia.org/wiki/List_of_Russian_milit ary_bases_abroad#Planned

warheads, such as submarine bases and nuclear weapons storage sites, that are not considered for the use of two warheads in Table 9.1

As outlined in Table 9.1, the primary targets are military bases.[2] We have subdivided these into navy, air force, and army bases. We omitted military or other satellites in orbit as a class of target since they could be easily destroyed with conventional weapons as already demonstrated by China and the United States. There are already satellites that are capable of maneuvering to other satellites and grabbing them and moving them to a different orbit. Possibly the maneuvering satellite could also destroy the other satellite. We also separated the military bases not within Russia or the United States but instead in allied countries. Navy bases and air force bases are generally tied to large facilities, such as harbors or airfields. However, army units can be small and widely dispersed. Therefore, many army units may be too small to be worth targeting. Air force units includes radars or air defenses that are not associated with airfields; some of these are also small. In some cases, ICBMs may be collocated with military bases. The US Base Structure Report (Department of Defense 2018) lists 4,150 bases in the United States, but most of these are small, not active, or National Guard facilities, which we assume are collocated with active Department of Defense bases or commercial airfields.

Another significant group of targets is large commercial airports.[3] In a war, military aircraft would likely be dispersed there. Oil storage units, oil refineries,[4] and power plants[5] would be important targets to stop the opposing country from being able to mount a ground assault or to stop its population from mobilizing after the initial attacks. In 2017, there were 536 refineries globally producing 79 million barrels of oil per day. Table 6.1 lists only large refineries producing more than 2,000 barrels of products per day. The refineries are also locations where fuel is stored in large quantities, which is important for evaluating the effects of fires. We have separated the types of power plants because bombing nuclear power plants has the potential to release large amounts of radiation. Also, bombing hydropower plants has the potential to create devastating floods. Although attacks on nuclear power plants, dikes, and dams are forbidden under Protocol 1 of the Geneva Convention, because neither Russia nor the United States has agreed to obey this treaty, we consider those targets. There are about 4580 power plants in the United States, but we assume only large ones, with design capacity above 1500 MW, would be attacked. A 1500 MW power plant can service about 600,000 houses. Finally, there are a group of lower-priority targets designed to

---

[2] U.S. military forces, https://data.opendatasoft.com Military bases; Russian Military forces, https://gfsis.org.ge/maps/russian-military-forces
[3] https://ourairports.com/airports.html
[4] https://www.fractracker.org/2017/12/global-oil-refineries-emissions/
[5] https://data.opendatasoft.com

stop the country from functioning. These include harbors[6] and railyards; government research centers focused on building nuclear weapons, space vehicles, or weather forecasting; and military-supporting industries that build such items as aircraft, missiles, tanks, or rifles. Russia would also be likely to explode a small number of weapons at high altitude over the United States to create an electromagnetic pulse (EMP). Such pulses occurred during nuclear weapons testing prior to the above-ground test ban in 1963. EMP would likely damage electrical networks in the United States and Canada as well as equipment such as cell phones, radios, and other equipment that was not hardened (Starr, 2023). Probably, Russia would not detonate such weapons over Europe to protect itself from EMP, and the United States possibly would not use them over Russia to protect NATO or Asian allies.

Instead of attacking ICBM sites, which are likely to be empty if a first strike occurred, nuclear weapons used in retaliation could be used against cities and other targets to make it more difficult for the adversary to recover. There are 50 US state capitals and the federal capital. The Russian Federation has about 85 federal subjects, which for historical reasons are variously called krays, oblasts, republics, and cities of federal importance. Two of these federal subjects are the city of Sevastopol and the Republic (or Autonomous Republic) of Crimea, but these two sites are not recognized by most of the world as being part of Russia. These are widely considered as counterforce targets and are already included in the target list in Table 6.1. There are also capitals of US allies such as Canada, Japan, and South Korea that would be possible targets for strategic weapons. The United States has about 300 cities and Russia has about 200 cities with more than 100,000 people. These might also be targeted instead of ICBM silos in retaliatory countervalue attacks. It is likely that warheads and delivery systems in storage of Russia or the United States would be destroyed in the initial attack so that subsequent nuclear conflict might not occur.

Russia may have about 1600 tactical nuclear weapons (Table 7.2). These are not deployed. Targets would include Europe since many nuclear weapons are located there either on US bases or in Britain and France. The NATO targets listed in Table 9.1 might be attacked by Russian tactical weapons. There may be 334 tactical weapons on bombers and air-to-ground missiles in the Russian Air Force, and 784 in the Navy, many of which are on submarine-launched cruise missiles according to Kristensen et al. (2024b). There are also ground-launched cruise missiles, which are in violation of the now defunct 1987 INF Treaty. Many tactical nuclear weapons might be used on the battlefield, such as torpedoes, mines, depth charges, and short-range weapons meant to destroy an advancing army, defend against air or missile attacks, and defend coastal

---

[6] https://publibrary.sec.usace.army.mil/ Search for "Principal Ports of the United States-2017"

areas. The yields of the Russian tactical weapons are not known but could be comparable to strategic weapons.

Table 9.1 lists many fewer weapons used on NATO than on Russia or the United States even though the population of the European Union is about as large as that of the United States and Russia combined. Partly this smaller number reflects uncertainty about the number of tactical weapons that Russia has that could be used quickly. In addition, Russia might want to minimize infrastructure damage to Europe so that they could occupy it after the war. Thus, they would attack large military bases. There are about 13 nuclear weapon storage sites in Europe that would be targeted. Nuclear attacks on oil refineries and power plants that might support tank attacks as well on commercial airfields that might support military aircraft in an emergency might be expected. It seems likely to us that Russia would not attack European assets such as railyards, commercial harbors, industries, and laboratories since they might use these after they invade Europe and to recover from attacks on Russia. Alternatively, one could also argue that Russia would attack these assets because they might be used by Europe to attack Russia.

Figures 9.1–9.4 illustrate the targets listed in Table 9.1. Figure 9.1 shows that US targets are widely distributed around the world. We have chosen only the 23 largest military bases outside the United States and NATO. Japan, South Korea, and Guam all have several such targets. There are also large military bases in the

Figure 9.1 Counterforce targets in the United States and among its non-NATO allies. There are 23 targets outside the United States. Red circles are military bases overseas and in the United States. The size of the circles does not represent the size of the damage zone from a nuclear explosion; they are sized for visibility on the map. Table 9.1 identifies the colors of other types of targets.
Map created with QGIS and OpenStreetMap

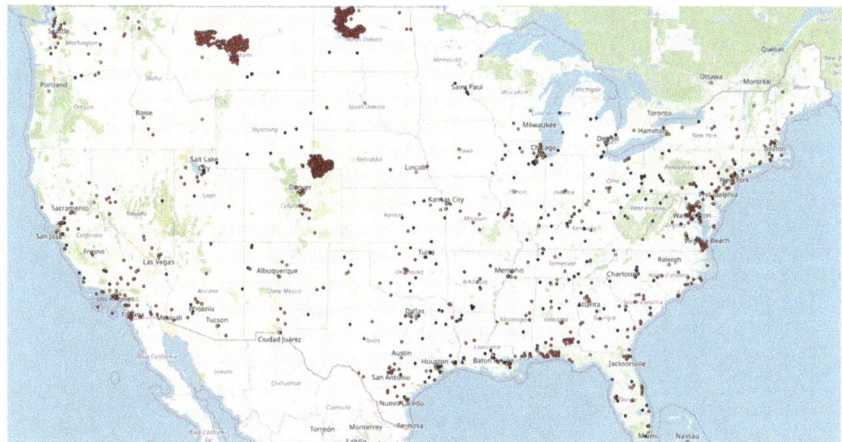

Figure 9.2 Counterforce targets in the contiguous United States. Every state, including Alaska and Hawaii, has more than one target. The size of the circles does not represent the size of the damage zone from a nuclear explosion; they are sized for visibility on the map. Circle colors are defined in Table 9.1.
Map created with QGIS and OpenStreetMap

Middle East. Australia has at least two important targets involving deep space communications and submarine communications systems. The circles on the graph show the locations of the targets, not the area that is damaged. In Chapter 10, we will focus on a few areas with multiple targets to show how the damage footprints overlap.

Figure 9.2 shows that each U.S. state has at least two targets.

Figure 9.3 illustrates targets in NATO countries (excluding the United States and Canada). Many of the NATO targets are also US military bases, and some store nuclear weapons. There are US nuclear weapons stored in underground vaults in Belgium, Germany, Italy, the Netherlands, and Turkey. In addition, France and the United Kingdom have nuclear warheads as well as submarines that are deployed with nuclear weapons.

Figure 9.4 illustrates targets in Russia and its allies. Targets are concentrated in the European part of Russia, and except for large airfields there are few that are found in the northern parts of Asian Russia. There are a number of targets in Russian allies, such as Armenia (3), Moldova (1), Abkhazia (mostly recognized as part of Georgia) (3), Tajikistan (2) Syria (2), Belarus (5), Ukraine (including Crimea) (30), and Kyrgyzstan (1). Ukraine in particular has a large number of targets due to Russian bases in the Crimea.

The targets of nuclear weapons are highly classified. Despite rules of war that ban attacks on cities, nuclear power plants, or dams, such rules could be easily

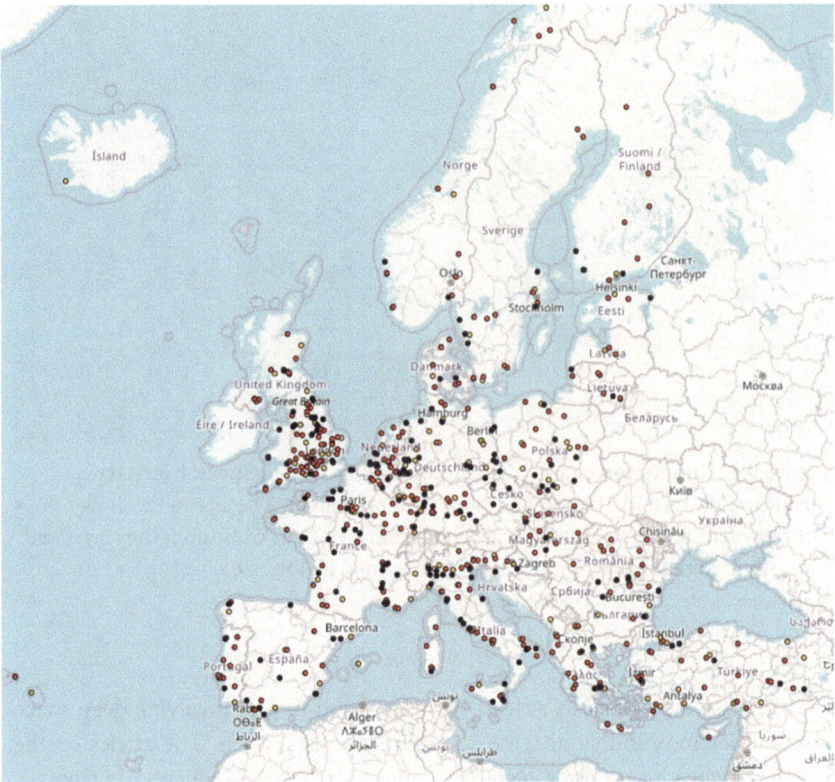

Figure 9.3 The map shows 638 targets in the European NATO countries. The size of the circles does not represent the size of the damage zone from a nuclear explosion; they are sized for visibility on the map. Circle colors are defined in Table 9.1.

Map created with QGIS and OpenStreetMap

ignored by claiming military targets are in cities, while dams and power plants are important to supply energy to the military. Counterforce targeting of military bases, large airfields, military-supporting industry, oil refineries, seats of government, and similar targets listed in Table 9.1 would almost certainly entail attacking many urban areas, though modern analyses of the overlap is lacking. There are no public reports by the US military since 1978 (OTA 1979) about the collateral damage to civilians in a nuclear conflict considering only assured destruction due to blast waves and nuclear radiation. Worse, it was recognized in the 1978 report that the military was underestimating civilian casualties because it ignores the effects of fires. These topics are the subjects of the next chapter.

Figure 9.4 The map shows 1,325 targets in Russia and its allies. Some military targets in the Arctic are off the map. The size of the circles does not represent the size of the damage zone from a nuclear explosion; they are sized for visibility on the map. Circle colors are defined in Table 9.1.

Map created with QGIS and OpenStreetMap

Chapter 10
Assured Destruction by Nuclear Explosions

"Assured destruction" of a target is destruction that the military has high confidence that it can predict and then achieve during an attack, or alternatively the amount of destruction of the target done in the least favorable circumstances for creating damage. In either definition, it is likely that assured destruction underestimates what might actually occur.

When a nuclear weapon explodes, it creates a shock wave capable of destroying reinforced concrete buildings, emits prompt nuclear radiation that can be fatal to those exposed, and makes a pulse of light so intense that it will burn exposed flesh and start fires if fuel is available. Except for nuclear radiation, these effects are similar to ones that occur after an asteroid or comet impact. The area over which these effects occur varies with the yield of the weapon and the height at which it is detonated, among other things. People can be shielded from prompt nuclear radiation and the light pulse by being in a building, and fires will only occur if there is material to burn. Assured destruction following a nuclear attack on a target is likely computed by the military based only on the shock waves since the effect of fires are uncertain according to Eden (2004).

In this chapter, we evaluate the mechanisms for assured destruction by nuclear explosions and estimate the casualties that might occur for various scenarios for nuclear explosions. Most of the effects discussed here can be evaluated using the web-based tool NUKEMAP,[1] which allows one to compute the areas damaged for any city and for various yields of nuclear weapons.

10.1 Assured Destruction by Shock Waves from Nuclear Explosions

A shock wave, or blast wave, is a pressure disturbance in the air that can be characterized by the pressure difference across the shock wave. The pressure difference is called the overpressure, which is measured relative to normal

---

[1] NUKEMAP, by Alex Wellerstein, https://nuclearsecrecy.com/nukemap/

Earth in Flames. Owen Brian Toon and Alan Robock, Oxford University Press. © Oxford University Press (2025). DOI: 10.1093/9780197799734.003.0010

atmospheric pressure, which at sea level is about 14.7 psi (pounds per square inch) or 1,013 hPa (hectopascals or millibars). Pressure is defined to be force per unit area.

Data from numerous weapons tests, as well as numerical simulations, have determined the magnitude of the overpressure that occurs for various sized explosions. Figure 10.1 illustrates how overpressure values of 1 psi, 5 psi, and 20 psi vary as a function of distance for airbursts with a range of yields. In all cases, an optimum burst height was chosen to maximize the area covered by the 20 psi overpressure. The area within which the shock wave overpressure exceeds a certain value can be maximized for a particular choice of yield by choosing an altitude for the detonation called the optimum height of burst.

The optimum height of burst to cover the largest area with a 20 psi or greater overpressure for a 1 Mt airburst is about 1820 m (1.1 miles). For Hiroshima, the overpressure at the surface at ground zero was probably slightly greater than 25 psi (1700 hPa), while for a 1 Mt airburst it is about 62 psi (4274 hPa). The overpressure for a 1 Mt groundburst is about 36,000 psi (250,000 hPa) at 0.1 km from ground zero and more than 100 million psi (700 GPa) at ground zero, which is about twice the pressure at the center of the Earth. The much larger overpressure for a groundburst than an airburst illustrates the value of groundbursts for destroying buried objects near ground zero. A 1 Mt groundburst will create a crater almost 400 m (1300 feet) in radius and 100 m (328 feet) deep. In contrast, the effects of an airburst extend farther from ground zero and encompass a larger area. So, airbursts are preferred for destroying large area targets such as a city, while groundbursts are used to attack buried targets.

Nuclear weapons tests and laboratory tests have determined the overpressure that will destroy various objects. For example, an overpressure of about 3000 psi may be needed to destroy a hardened missile silo, but 20 psi is enough to destroy a reinforced concrete building. A 20 psi pressure difference is about 30% greater than normal sea level pressure, so based on common experience it may seem surprising that it can destroy buildings. We do not experience a pressure difference standing in the air since the pressure is the same in all directions. Or put another way, if atmospheric pressure pushed down on your hand and was not balanced by atmospheric pressure pushing up from the bottom of your hand, the average male hand would hold about 225 kg and brick structures out to 1.5 km where (500 pounds). Ouch!

The susceptibility of buildings to damage may depend on many factors, including construction materials, construction methods, orientation to the blast wave, and other factors according to Glasstone and Dolan (1977). Data from Hiroshima and above-ground nuclear tests on the effects of shock waves show that glass windows shatter for peak overpressures of 0.5–1 psi, much as they did

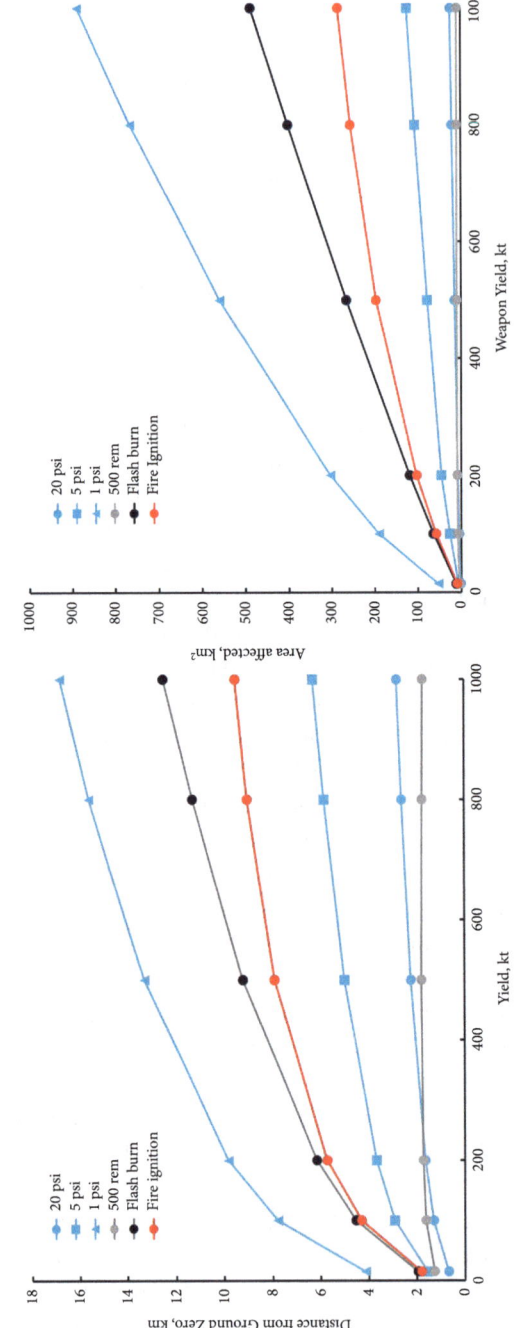

Figure 10.1 The distance from ground zero (left) and the area within (right) overpressure values of 1, 5, and 20 psi; the 500 rem radiation level, the light levels needed for flash burns, and the light levels needed for fire ignitions based on Glasstone and Dolan (1977) and NUKEMAP.

in the Chelyabinsk meteor shock wave. Unreinforced brick, wood, cinderblock, or concrete wall panels will shatter in the range from 1 to 10 psi. Above 10–20 psi, most buildings, including reinforced concrete, will be destroyed.

Overpressures near 50 psi are needed to kill people (White et al. 1960). Overpressures slightly larger than 25 psi occurred near ground zero in Hiroshima. While reflections and focusing increased the overpressure locally, it is likely that most people were not killed by the overpressure directly. Overpressures were sufficient to blow down reinforced concrete buildings in Hiroshima, a 15 kt explosion, at about 300 m (1000 feet)—1 km (3300 feet) from ground zero as shown by the locations of the 20 and 10 psi values in Figure 10.1 and wood and brick structures out to 1.5 km where the overpressure was about 5 psi, as shown in Figure 10.1. Fatalities may approach 100% when multistory buildings collapse on top of people.

In addition to the overpressure in the shock wave, there are also strong winds associated with the shock waves that can have velocities of hundreds or thousands of km/hr (hundreds or thousands of miles per hour). Category 5 hurricane force winds (157 miles per hour, 253 km/hr) extended more than 1.5 km from ground zero at Hiroshima and lasted for a few seconds. These winds were likely the main destructive aspect of the blast wave near ground zero for people who were outside buildings at Hiroshima. Exposed people would have been hurled through the air at high speed, while many people would have been hit by flying debris.

Figures 10.2a,c and 10.3a,c illustrate for 100 kt and 500 kt airbursts the areas impacted by shock waves within the 20 psi contour and the 5 psi contour for attacks on targets in St. Petersburg, Russia, and San Diego, California, using the target lists from Chapter 9. Twenty psi shock waves will destroy reinforced concrete buildings, and 5 psi will destroy typical residential buildings, including brick. Figures 10.2b,d and 10.3b,d show the areas where fires could potentially be ignited by the 10 calorie/cm^2 flash of light as discussed below. San Diego and St. Petersburg are highly likely targets in a nuclear war. Each city has extensive harbors. San Diego is a base for aircraft carriers and has several military airfields. Its population in 2024 was 1.4 million. It is an important cultural center with a world-famous zoo. St. Petersburg has several important shipyards as well as military-supporting industries and military headquarters. Its population in 2020 was 5.6 million, making it the fourth most populous city in Europe and the second in Russia. It is a center for Russian culture, with world-renowned museums and palaces.

Figures 10.2a, c and 10.3a, c illustrate the much larger areas impacted by the blast waves from 500 kt yield weapons compared with 100 kt weapons. However, the 100 kt areas including fires (Figs. 10.2b, 10.3b) are much larger than the 500 kt areas considering only 20 psi shock waves (Figs. 10.2c, 10.3c). While both the 100 and 500 kt shock waves of 20 psi produce somewhat isolated damage areas, the 500 kt (Figs 10.2d, 10.3d) fires basically obliterate both cities.

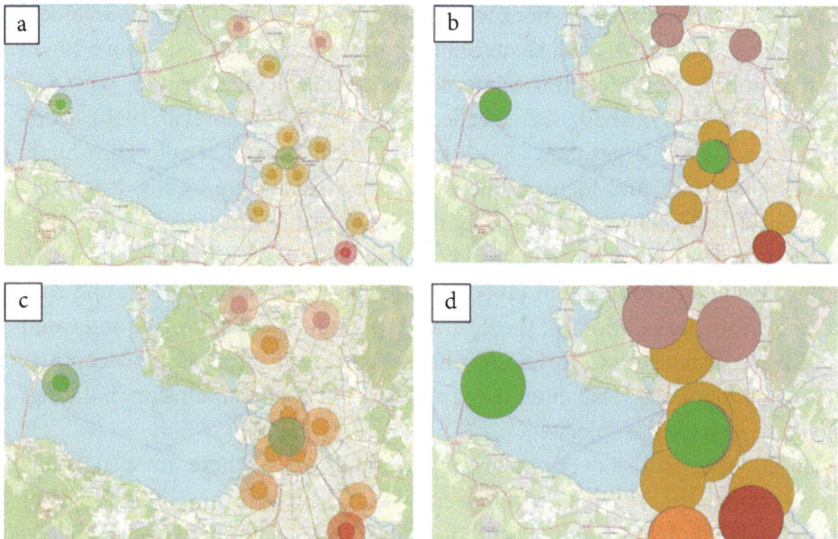

Figure 10.2 Contours of destruction for an attack on St. Petersburg, Russia: a) 20 psi contours and 5 psi contours (outer circle) for 100 kt weapons, b) 10 cal cm^{-2} contour for 100 kt weapons, c) and d) are the same as a) b) except for 500 kt weapons. Commercial airports are orange, navy bases are green, arms makers are mustard, power plants are red, and military airfields are pink.

10.2 Assured Destruction by Prompt Nuclear Radiation

Radiation is not as simple to evaluate as blast damage because people can be sheltered from it simply by being inside of large buildings. Also, deaths may take a long time and depend on healthcare for moderate exposures. However, radiation is certain to occur. There are several different types of radioactivity created in a nuclear explosion. Generally, these are divided by the time span over which the radiation is produced. We will return to local fallout and global fallout later in this chapter. Here we discuss the prompt nuclear radiation.

Sidebar 10.1 Units of Radiation

Different types of radiation have different impacts on the body. There are numerous units of radiation dose that quantify various aspects of the radiation. The rem unit is a dose equivalent unit designed to make different types of radioactivity equivalent in terms of damage to the body. The lethal dose, LD_{50}, of radiation expected to cause death to 50% of those exposed within 30 days is 400–450 rem received over a period of days or less. Such a short time exposure is called an acute dose. The same rem received over a long period of time, such as a human lifetime, may have little impact since the body can repair some damage if it occurs slowly. Many people use the more modern unit of a sievert (Si). One hundred rem is the same as 1 sievert.

A rem is the dosage of ionizing radiation that will cause the same biological effect as 1 roentgen of X-ray or gamma-ray exposure. A roentgen of gamma rays will ionize about 5 out of every billion molecules of air or produce about 1.6×10^{15} ions per kg of air at normal temperature and pressure. Ionizing radiation is radiation that has sufficient energy to remove electrons from atoms and thereby make the atoms electrically charged, or to destroy the DNA in cells. Particles that are ionizing generally travel faster than 1% of the speed of light. Light that is ionizing has a smaller wavelength, or higher frequency, than most ultraviolet light, and therefore has the energy to knock electrons off atoms or destroy DNA in cells. Neutrons traveling at a few percent of the speed of light are more damaging to the body per unit of energy carried than gamma rays, which are a wavelength range of light. Therefore, fewer neutrons need to be absorbed by the body to produce the same rem as gamma rays.

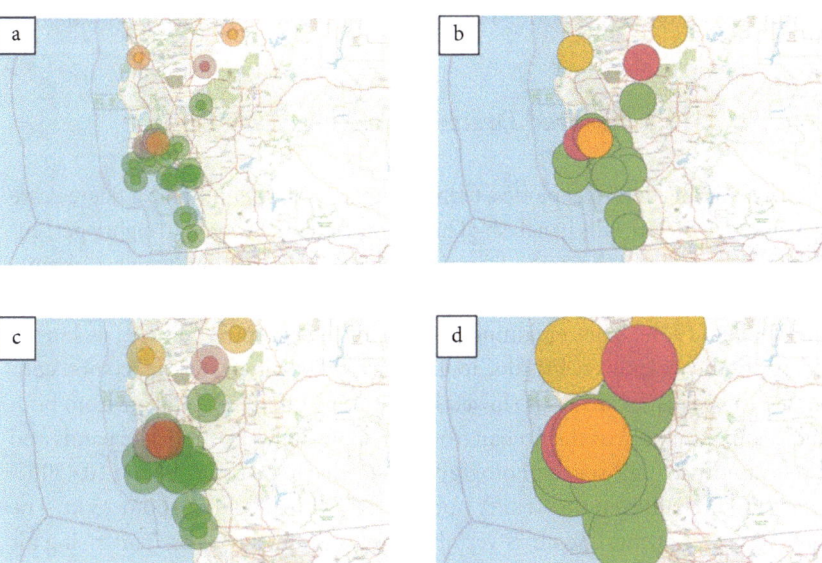

Figure 10.3 Contours of destruction for an attack on San Diego, California, United States: a) 20 psi contours and 5 psi contours (outer circle) for 100 kt weapons, b) 10 cal cm^{-2} contour for 100 kt weapons, and c) and d) are the same as a) b) except for 500 kt weapons. Commercial airports are orange, navy bases are green, arms makers are mustard, power plants are red, and military airfields are pink.

Chapter 6 illustrated a few of the possible nuclear chain reactions in fission bombs. These chain reactions result in the production of fragments of the original fuel for the bomb, many of which are traveling at speeds that are a few percent of the speed of light. There are many unstable isotopes, fragments of atoms, as well as neutrons, and gamma rays produced in fission bombs. Fusion reactions also produce gamma rays, but most of the radiation from thermonuclear bombs comes from fission. Many of the isotopes and atomic fragments from nuclear reactions are quickly stopped by the air. However, the neutrons and gamma rays can travel moderate distances.

As indicated in Figure 10.1, for the Hiroshima bomb the greater than 500 rem prompt nuclear radiation exposure area, where 50% of exposed people may die, extends to about 1 km from ground zero. Therefore, the prompt nuclear radiation hazard is most important near ground zero. The area exposed to 500 rems or more is nearly independent of the fission yield of the bomb. This independence occurs because the atmosphere very efficiently stops the high energy particles and light over kilometer distances.

Prompt nuclear radiation is not efficiently blocked by buildings. The typical wooden Japanese house at Hiroshima was found to experience about 60–90% as much gamma radiation inside the house as outside the building, with the amount of exposure being reduced if one is further inside the walls of the building. Even buildings with concrete 30 cm (1 foot) thick only reduced the amount gamma radiation by a factor of 4 relative to the amount outside the building.

10.3 Assured Destruction from Flash Burns

When an atomic bomb explodes, temperatures of tens of millions of degrees are generated, which results in a flash of high-energy light. Seventy to eighty percent of the energy of the explosion is in the form of this radiation, which is mainly X-rays that are absorbed by the air near the explosion site according to Glasstone and Dolan (1977). The remaining 20–30% of the energy is released as kinetic energy of the weapons debris due to its velocity. The kinetic energy is soon converted to heat as the debris is slowed by the atmosphere. The hot air from both these sources is seen as the fireball from the explosion. The fireball expands and cools rapidly so high-speed photography is needed to record it. Figure 10.4 illustrates the fireball from the Trinity nuclear test, with a yield of about 25 kt. The fireball is so hot that the air is very buoyant and quickly rises. This rising air creates the familiar mushroom cloud, illustrated for the Hiroshima explosion by the photograph shown in Figure 10.5. Photographic analysis suggests the top of this cloud was near 9 km. The white column is mostly water vapor from the atmosphere condensing in the rising and cooling air. At the base is black smoke and debris rising from the destroyed city.

During its first seconds, the fireball has a temperature like that of the Sun, in the range of 6,000 K (10,340°F), and radiates light at visible wavelengths. An intense burst of visible light is created that can be blinding to those nearby and cause severe burns. The duration of the light pulse is short, a few seconds depending on the yield, but infrared radiation continues to be emitted by the fireball in the first minute or less of the explosion. Some of the energy in the fireball creates shock waves, but 35–45% of the energy of the explosion is contained in the visible and infrared radiation.

The radiation is emitted in all directions, and its energy density falls off in proportion to the square of the distance from the fireball. For blasts with yields of 1 Mt or less, the loss due atmospheric transmission is not significant unless the atmosphere is foggy or smoke-filled.

The light energy needed to create third-degree burns (the most serious) in 100% of the exposed population ranges from 8.7 calories/cm² for a 15 kt weapon

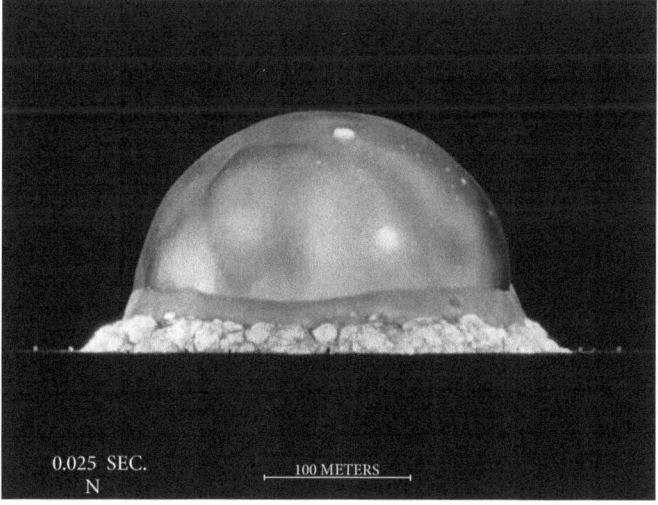

0.025 SEC.
N
100 METERS

Figure 10.4 The Trinity test of the first plutonium atomic bomb, 0.025 seconds after detonation, at Alamogordo, New Mexico, July 16, 1945. The fireball appeared red and yellow to the observer, but no high-quality images exist in color. Upon witnessing this, J. Robert Oppenheimer, director of the Manhattan Project, quoted the *Bhagavad-Gita*, "I am become death, the destroyer of worlds."

Figure 10.5 The mushroom cloud over Hiroshima a few minutes after the explosion on August 6, 1945, of a 15 kt atomic bomb. The photo was taken by Bob Caron, tail gunner of the Enola Gay, the airplane from which the Americans dropped the bomb. The top of the cloud was above the 9,144 m (30,000 ft) flight level of the Enola Gay at this time.
National Archive

to 11.3 calories/cm^2 for a 1000 kt weapon according to Glasstone and Dolan (1977). The energy needed for third-degree burns is a function of yield because for higher-yield weapons the light pulse lasts longer and the energy has more chance to propagate into the target. These data are uncertain since different skin colors will absorb more or less of the radiation. The fatality rates for third-degree burns depend on many factors; most important are the area of the body covered by the burns and the age of the person. Third-degree burns are potentially fatal if they cover more than 20% of the body and if no medical care is available.

Sheltering from thermal flash burns can be provided by even a wooden house, although flash burns did occur in Hiroshima for people standing near windows. At lower light levels, thick clothing might provide some shelter against flash burns, but such clothing might also burst into flames at higher light levels.

10.4 Destruction from Fires

The effective thermal radiation that does reach the surface only lasts for seconds. It chars thick surfaces such as the siding on buildings but does not lead to flames. However, thinner materials such as cotton clothing, window shades, upholstery in cars or on furniture, newspapers and other papers, leaves, rotted wood, grass, pine needles, and similar materials ignite when exposed to light levels in the range from 5–15 calorie/cm^2 according to Glasstone and Dolan (1977).

Easily ignitable fuels within the distance from ground zero at which more than 10 calorie/cm^2 of radiant energy is deposited are likely to catch fire after a nuclear explosion. Since the light pulse travels faster than the shock waves, it is possible for the shock waves to blow out the fires or bury them in rubble. The fires may start simultaneously in many individual locations where suitable fuels are available. For instance, plant debris outside of buildings and materials inside buildings exposed to the light such as curtains in windows are likely to start on fire. These small fires are likely to spread within a building or in vegetation outside the building. Even if fire departments were not destroyed in the nuclear attack, and firefighters were not prevented from entering the fire area due to debris blocking roads, the sheer number of fires would make it difficult or impossible to extinguish them. In Hiroshima, it took about 20 minutes for a firestorm to develop. Winds reached 45–60 km (30–40 miles) per hour two to three hours after the explosion, suggesting the fire intensity peaked about then. A pyrocumulonimbus (pyroCb), a thunderstorm started by a fire, formed over the region as convective motions carried smoke and humidity about 16 km (10 miles) above the city (Figure 10.6) (Baba et al. 2011). The winds decreased after 6 hours, suggesting the fire intensity diminished at that time. Due to the protracted duration of the fires, some people who were not seriously injured were able to escape from the burning area.

As Figures 10.1, 10.2, and 10.3 illustrate, the region over which fires may be ignited greatly exceeds the area within the 20 psi shock wave. The strong

Figure 10.6 The smoke cloud from the Hiroshima firestorm generated by the fires ignited by the bomb. Taken more than 3 hours after the bomb blast on August 6, 1945. The top of the cloud was estimated to be at 16 km (10 miles). This photo is often mistaken for the mushroom cloud from the nuclear explosion.

shock waves only cover 10–15% of the area that can be ignited. In Hiroshima and Nagasaki, there is direct evidence of the power of shock waves in the collapse of buildings of various types. For airbursts, the shock waves generally propagate downward, and consequently they tend to destroy roofs if they are not powerful enough to collapse the building. However, there is also evidence that most of the combustible material burned. Some collapsed buildings buried flammable material, protecting it from burning. However, many buildings that did not collapse were destroyed inside by fire, and some collapsed buildings exposed flammable material to flames. In parts of Hiroshima, people were able to flee beyond the fires before they spread to the entire region within about 2 km of ground zero. Essentially, no buildings were burned out beyond 2.4 km (1.5 miles) from ground zero, while 50% were burned out 1.6 km (1 mile) from ground zero.

10.5 Casualties and Fatalities in a City from Explosions of Nuclear Weapons

The nuclear explosions at Hiroshima and Nagasaki provide our only direct evidence for casualties and fatalities from nuclear blasts. There are many uncertainties about the population of these cities before the attack, and esti-mates of the numbers who were killed or injured also vary considerably, as noted by Wellerstein (2020). However, estimates of the fraction of the popula-tion killed or injured as a function of distance from ground zero are less variable. Figure 10.7 shows data from Hiroshima on the probability of death or casu-alties as a function of distance from ground zero as colored lines (Ishikawa and Swain 1981; Oughterson and Warren 1956). Casualties include fatalities plus those severely injured, so the difference between the curves is the frac-tion that are injured but not killed. As the figure shows, near ground zero nearly everyone was killed. Data from Nagasaki show very similar values to those from Hiroshima. Figure 10.7 indicates that about 38% of the people within 1.5 km of ground zero died, and about 48% either died or were seriously injured.

The highest fatality probability is near ground zero. Figure 10.8 shows that in Hiroshima, like most cites, the population density declined away from the city center. However, the area in a ring around ground zero increases with distance. Therefore, as shown in Figure 10.8, for Hiroshima the peak of the fatalities per

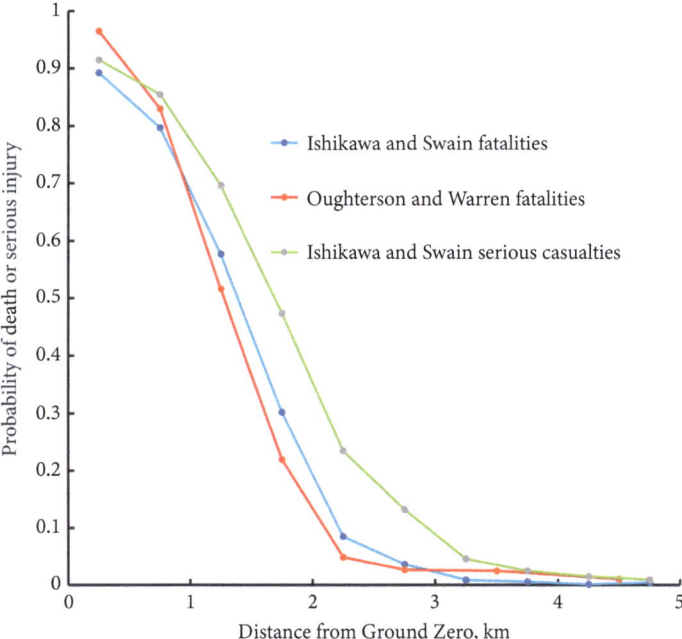

Figure 10.7 The probability of death or serious casualty as a function of distance from ground zero. The observed deaths in Hiroshima are given from two data sources in the red and blue curves. The observed serious casualties are given by the green curve. Serious casualties are the sum of fatalities and serious injuries.

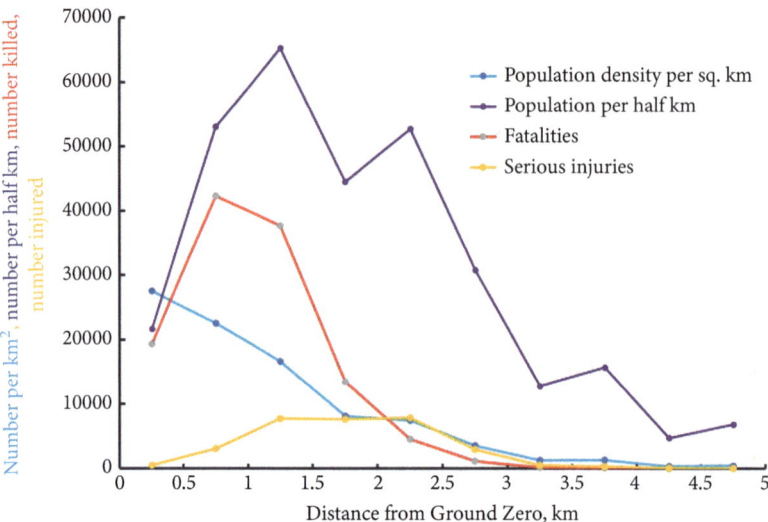

Figure 10.8 To determine fatalities per half km, the population needs to be multiplied by the probability of death shown in Figure 10.7.

half km occurred between 0.5 and 1.5 km from ground zero. Serious injuries per half km peaked farther out, between 1.5 and 2.5 km from ground zero. Most people closer to ground zero than 1.5 km died instead of being injured.

Glasstone and Dolan summarized the various factors that caused fatalities in Japan. Since more than one potentially fatal factor occurred, it is difficult to assign deaths to an individual factor. About 50% of all deaths were caused by various types of burns. About 30% of those who died received lethal doses of radiation but may have actually died from burns or other factors such as building collapse, because death from radiation is relatively slow. It is thought that 5–15% of deaths were due to prompt nuclear radiation. Flash burns accounted for 20–30% of the fatalities in Hiroshima and Nagasaki. Mechanical injuries can come from many sources. About 50% of people directly exposed to an overpressure of 50–75 psi will die, but these overpressures only occur for 1 Mt airbursts near ground zero and did not occur at Hiroshima or Nagasaki. Shock waves may reflect and amplify. Many are killed by flying objects, being blown into walls and other solid objects, or collapse of buildings due to overpressure.

While it is difficult to evaluate the biological responses to the various physical phenomena that occur following a nuclear explosion, the area within which the overpressure exceeds 5 psi, the thermal pulse exceeds 10 calorie/cm^2, and the prompt nuclear radiation exceeds 450 rems can all be computed for any

yield and height of burst. Therefore, we can extrapolate the fatality and casualty curves for Hiroshima to other nuclear explosions using the physical factors
for overpressure, thermal pulse, and prompt nuclear radiation, rather than the
biological factors related to survival. It is important to note that thermal radiation refers to the light intensity from the explosion that is sufficient to start fires.
Flash burns refers to the light intensity that is sufficient to burn exposed skin.
Prompt nuclear radiation is the intensity of gamma rays and fast neutrons that
is sufficient to cause death.

Figure 10.9 shows our estimate of the probabilities of death from prompt
nuclear radioactivity, thermal radiation, shock waves, and fires and their total
for Hiroshima. Figure 10.10 shows the probability of death for a 1 Mt airburst
at 1820 m above the ground. A 10% chance of death occurs at about 2.25 km
from ground zero for Hiroshima and 8.5 km for a 1 Mt explosion. Based on
data from survivors, we assume there was a 40% chance of being outside at
the time of the explosion at Hiroshima. At Hiroshima, prompt radiation, blast,
and flash burns were all of similar importance within 1 km of ground zero,

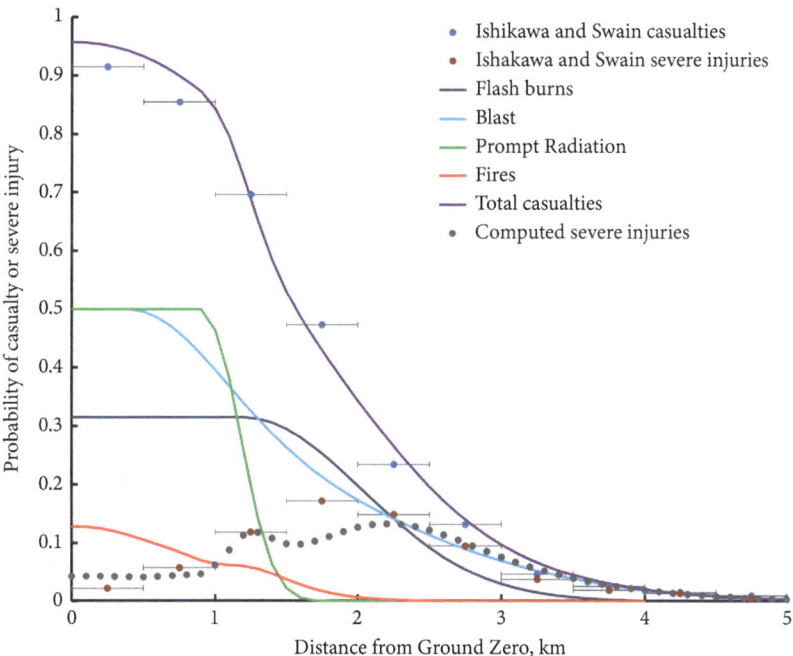

Figure 10.9 The probability of casualty from various phenomena that
contribute to the total for Hiroshima

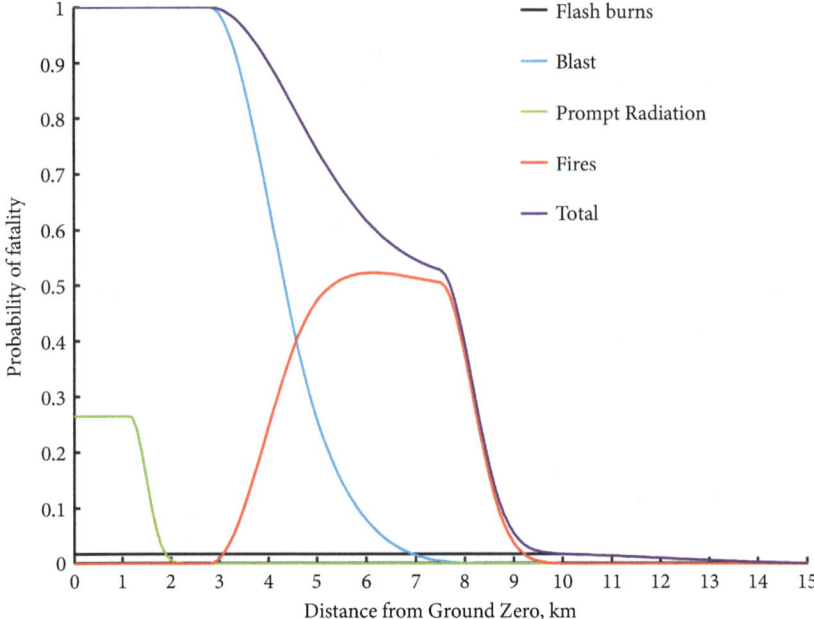

Figure 10.10 The probability of fatality for various phenomena that contribute to the total for a 1 Mt airburst

but farther out blast and flash burns dominated. Despite the firestorm, fires were not too important because those who were not injured could flee from the fire. For a 1 Mt nuclear explosion, blast dominates the fatalities for the first 4 km, but then fires become of equal or greater importance. Flash burns are not important because in modern cities fewer than 10% of people are outside on a normal day.

10.6 Deaths from Fallout

We have just discussed causes of death very soon after the nuclear explosions. However, radiation can lead to deaths long after the explosions. People are often most frightened of nuclear explosions because of the invisible threat from radioactivity. We have already described the potential casualties from prompt nuclear radiation emitted in the first minute from a nuclear explosion. There are four other sources of radiation that also pose problems.

10.6.1 Radiation from Long-Lived Isotopes

One issue in all types of nuclear explosions is the release of long-lived isotopes such as cesium 137, ^{137}Cs, with a half-life of about 30 years, and strontium

90, ^{90}Sr, which has a half-life near 29 years. It was concern about this type of radiation building up across the Earth that led to the 1963 ban on above-ground nuclear tests.

Sidebar 10.2 Natural Radiation Dose

It is estimated that the worldwide average annual effective dose of radiation received by the typical person from natural background radiation is about 2.4 mSv (0.24 rem), with a range between 1 and 10 mSv (UNSCEAR 1993). A milliSievert (mSv, one-thousandth of a sievert) is a unit of biological response to radiation. A sievert has units of joule/kilogram and represents the biological effect of the deposit of a joule of radiation energy in a kilogram of human tissue. A rem and a Sv have identical meanings with different units. A Sv is meant to represent an equivalent or effective dose, so weighting factors are needed to account for the type of radiation, the mode of exposure, and the organs exposed. The implications of a 1 mSv dose depend on whether the dose is gradual or acute (sudden) since the body can repair some radiation damage from a gradual dose. Generally, 1 mSv is taken as the limit of acceptable annual artificial exposure of individuals, since that dose is smaller than the exposure to natural radiation from cosmic rays, radon, and other natural sources. People in numerous occupations such as medical workers, air travel crews, and miners are exposed to several times higher doses than 1 mSv.

We can estimate the impact of long-lived radiation from a war between NATO and Russia by comparing to the radiation exposures from atmospheric nuclear weapons testing. Up to 1971, about 440 Mt of yield were exploded in the atmosphere, of which about 189 Mt of the yield was from fission in 543 tests (UNSCEAR 2000). The amount of radioactive material released in an explosion is proportional to the amount of the yield due to fission reactions. Table 7.2 estimates that the United States has about 287 Mt of deployed strategic nuclear warheads and that Russia has about 358 Mt of deployed strategic nuclear warheads. The total yield, assuming all the deployed weapons detonated, is about 645 Mt. The fission fraction of these weapons is approximately one-half, or 322 Mt. Hence the amount of radiation released in a war with the current deployed weapons is a little less than twice that exploded during atmospheric testing. If the stored weapons were also used, it is likely that about double the yield of the deployed weapons alone would be exploded.

The global average radiation dose for each person in the year 2000 from atmospheric testing is estimated to have been 0.005 mSv, while it was 0.14 mSv in 1963 (UNSCEAR 2000). The considerable reduction in dose from 1963 to 2000, which is much larger than from the decay of the radioactive elements, is because a lot of the material has migrated into the soil or washed into the sea where people are no longer so exposed to it. If we scale by the amount of yield from the weapons tests to the deployed arsenal, a nuclear war using the deployed strategic weapons might produce a dose of 0.24 mSv per annum in the years just after the war. If the weapons in storage were also used, a dose of 0.48 mSv per year from long-lived radiation might occur soon after the war. These doses are smaller than those received every year in many occupations such as by airline crew members. While additional cases of cancer would be expected, they would be difficult to detect from those resulting from natural sources of radiation.

10.6.2   Radiation from Ground Bursts

Sidebar 10.3   Yet More Radiation Units

A rad is a measure of absorbed dose of radiation; it represents the absorption of 100 ergs (10^{-7} joules) of nuclear (or ionizing) radiation per gram of absorbing material, such as body tissue. A modern equivalent unit is the gray. A gray (= 100 rad) is the amount of radiation produced, as opposed to a rem or a Sv, which are dose-equivalent amounts of radiation that attempt to equalize exposures, considering the parts of the body exposed, whether the radiation is external or internal, and the specific type of radiation released.

The best known, and most dangerous, source of radiation is from ground bursts. Ground bursts lift soil from the surface, to which the radioactive isotopes quickly become attached. Soil particles are large and are quickly deposited on the ground. Studies of this fallout from above-ground nuclear tests have allowed calculations of this hazard for weapons of various yields. However, where this radiation lands depends on the local wind speed and direction, which vary with the weather. Moreover, people can shield themselves from this radiation. Generally, the radiation is most hazardous for only a few days, and staying inside in a location where the dusty air cannot enter provides protection. Nevertheless, many evaluations of assured destruction find tens of millions of fatalities due to radiation from ground bursts.

The complexity of the nuclear debris makes calculation of the radiation exposure difficult because each isotope has a different lifetime, and the relative abundance of the isotopes therefore shifts with time. In addition, the decay of some of the isotopes can induce surrounding materials to become radioactive.

About 3×10^{23} fission product atoms are formed per kiloton of fission energy yield (Glasstone and Dolan 1977). The mass of this material is small, about 57 kg (125 lbs) per megaton of fission energy yield. It is observed that the ionizing radiation released by the sum of all the isotopes decaying declines with time for the first 6 months following the expression:

$$R = R_0/t^{1.2}. \tag{7.1}$$

In this equation, R is the radiation dose rate, in units of rads per hour, at time t (in hours) after the explosion, and R_0 is the dose rate that would have occurred after 1 hour had all the fallout been deposited at that location by that time. Hence, if the hypothetical dose from the radiative material after 1 hour is known, then the dose 7 hours later will be about one-tenth of that initial dose, after 2 days it will be about 1% of the initial dose, and after 2 weeks it will be about 0.1% of the initial dose (Glasstone and Dolan 1977). This simple relationship is accurate to within about 25% for the first 6 months following the explosion. However, the calculation of the dose is complicated by the fact that most of the radiation would still be in the atmosphere at 1 hour.

One can also compute the total dose received, R_{total}, at any time after the radioactive material stops falling. Here the units of the total dose are grays. Because the radiation decays in time, most of the dose is received just after the deposition of the radiation.

Once the decay rate and the dose are known, it becomes possible to compute the amount of radiation falling on the ground downwind of the explosion. Unfortunately, this calculation is complicated by the need to know the wind speed as a function of altitude, distance, and time, as well as the fall velocity of the dust particles and whether there is any rain along the path of the dust that might remove it from the air. Glasstone and Dolan (1977) provide several approximate treatments of this problem, and the reader can use web-based tools such as NUKEMAP by Alex Wellerstein to illustrate the radioactivity contours around any target area they choose.

Figure 10.11 shows the exposure contour for a 15 Mt ground burst at Bikini Atoll in the Marshall Islands in 1954, which Glasstone and Dolan (1977) use to illustrate an actual set of doses for an exposure extending to 96 hours after the explosion. There were 67 nuclear tests in the Marshall Islands, and their impact still lingers in requests for additional compensation for those sickened or killed by radiation and for those displaced from their homes by lingering radiation. The Castle Bravo explosion, the sixth largest ever conducted and the largest conducted by the United States, was notorious because the radiation reached inhabited islands and a Japanese fishing boat. The yield was almost three times larger than expected, and large dust particles fell out of the stratosphere more

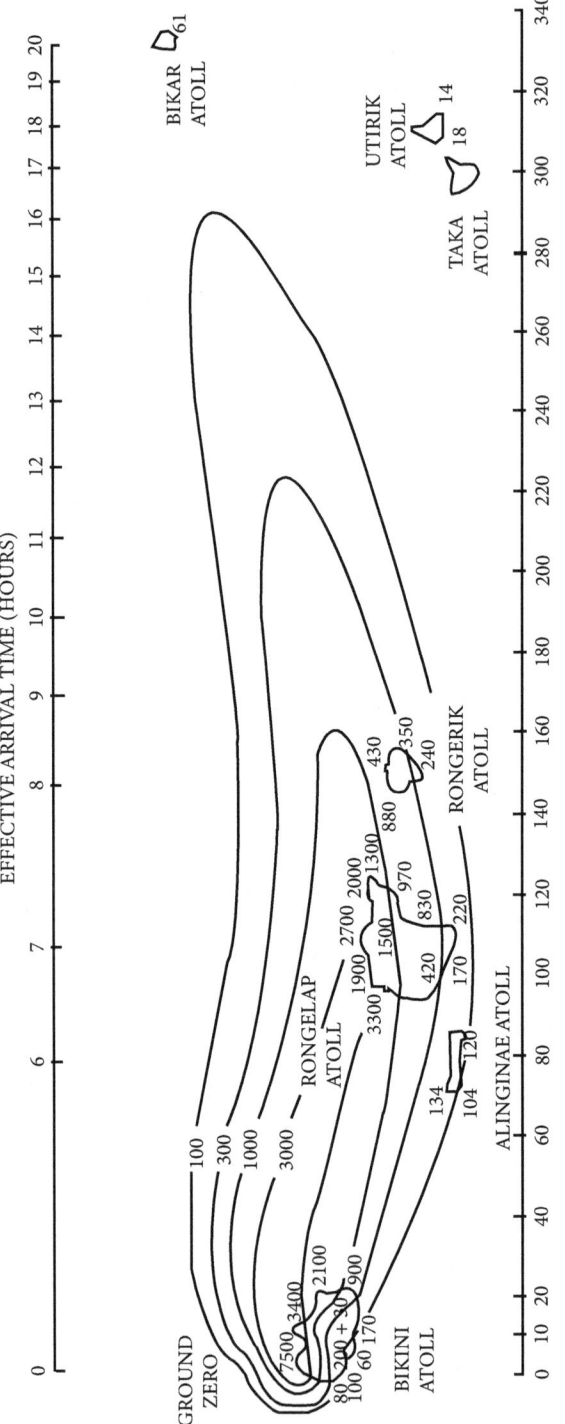

Figure 10.11 The estimated total accumulated dose contours in rads at 96 hours after the Bravo test explosion (Glasstone and Dolan 1977). The Bravo test was a 15 Mt explosion on the Marshall Islands on March 1, 1954. The explosion occurred about 2 m (7 ft) above the coral reef and was estimated to have contaminated about 18,000 square kilometers (7000 square miles) to such an extent that evacuation of the area would be needed to avoid death. There are little data from the Pacific Ocean on the actual exposures over much of the area, so models are not well constrained.

rapidly than expected. Evacuations from the islands took place three days after the explosion, and numerous inhabitants of the islands as well as those on the fishing boat experienced radiation sickness, and one fisherman died. The Rongelap Islanders absorbed a lot of radioactive iodine via their drinking water, and almost all the children had to have their thyroid removed. Considerable controversy remains about this event, due in part to uncertainty in the winds, and more modern versions of Figure 10.11 show an even greater area within the 100 rad contour (Kunkle and Ristet 2013).

An example of the radiation from an attack on the US ICBMs is shown in Figure 10.12. In this attack (Philippe 2023), the 450 silos in the United States were each attacked with one 800 kt warhead on various days in 2021. As Figure 10.12 shows, the location of the deposited radiation varies greatly from day to day with the changing winds and rainfall. Estimates of fatalities range from 340,000 to 4.6 million depending on the day, with an average of 1.4 million. Another analysis was conducted by Helfand et al. (2002) for the year 2000, when there were 100 more silos than in 2022. In this case, the people living in an area of about 400,000 km^2 (154,000 square miles) who were not sheltered would have received a 48-hour dose exceeding 4.5 Sv, the dose expected to be lethal to 50% of those exposed. Helfand et al. predict that 1,285,000 people could have died from this attack on missile silos if they were all unsheltered and about 400,000 if they all sheltered in a typical single-story residential building.

10.6.3 Radiation from Airbursts

Most weapons used in a nuclear exchange are likely to be airbursts because their destructive shock waves propagate farther than those from groundbursts. Therefore, airbursts do damage over a larger area than groundbursts.

Generally, studies of radiation exposure ignore the radiation from airbursts because older arsenals used in nuclear testing contained high-yield weapons whose fireballs rose far into the stratosphere. The radioactivity in these fireballs had no large particles to attach to and mainly attached to the very small-sized remains of the bomb casing. Consequently, the radiation from the fireball lingered in the stratosphere where it never rains and did not reach the surface for several years when it had largely decayed.

However, there is some radioactive material in the stem of the fireball, which is located in the lower atmosphere, and modern weapons have lower yields and so some of their fireballs do not rise completely into the stratosphere. This radiation could be deposited on the ground by rainfall. In the lower atmosphere, rain often occurs within a few days. Radioactive rainfall did occur after the Hiroshima airburst and did harm people on the ground.

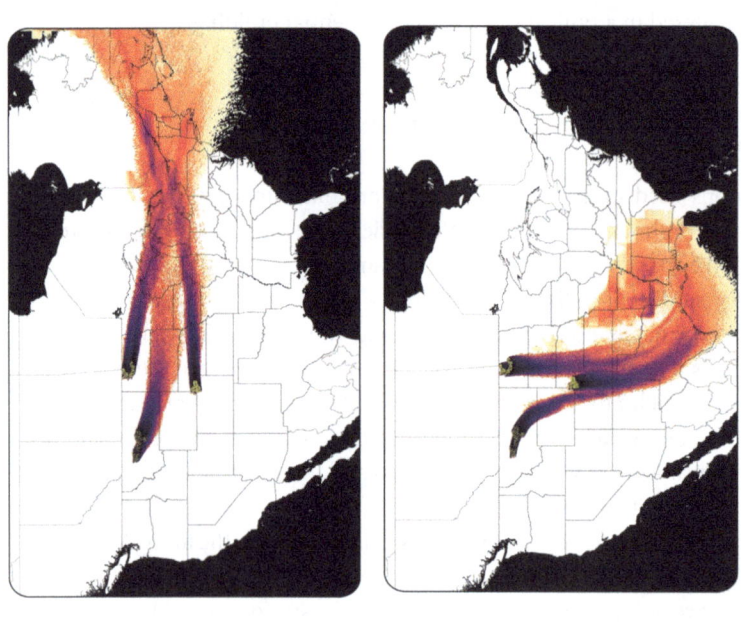

Cumulative Radiation Dose after 4 Days, in Grays

One gray (Gy) is equal to an absorbed dose of one joule per kilogram

- 84 Gy —— Death occurs within days of exposure
- 24 —— Death occurs within weeks of exposure
- 8 —— Death occurs 100% of the time
- 4 —— Death occurs 50% of the time
- 2 —— Death occurs 50% of the time for at-risk population
- 1 —— Death from acute radiation syndrome begins to occur in exposed population
- 0.5 —— Symptoms such as nausea and vomiting begin
- 0.1
- 0.05 —— Annual limit for radiation workers
- 0.001 —— Annual limit for the public

Figure 10.12 The cumulative radiation dose after 4 days from attacks on US missile silos on January 6, 2021, and February 24, 2021. Figures produced by Sébastien Philippe, Svitlana Lavrenchuk, and Ivan Stepanov. Reproduced with permission. © 2023 Scientific American, a Division of Springer Nature America, Inc. All rights reserved.

Unfortunately, it is difficult to quantify this threat from airbursts. The radioactivity decays very rapidly, so to pose a significant threat the debris must encounter rainfall within a day or two of the atomic blast. On April 27, 1953, radioactive debris from rainfall the previous night was detected in Troy, New York, at levels far above normal. This debris was found to be from a 43 kt nuclear test two days earlier in Nevada. The fireball rose to 9–12 km (30,000–40,000 ft) altitude, where it encountered the jet stream and was quickly carried to New York, where a violent thunderstorm deposited it on the ground.

We are not aware of any calculations of the radiation hazard from airbursts in nuclear wars.

10.6.4 Radiation from Nuclear Power Plants and from Stored Nuclear Waste

There is debate about whether a nuclear bomb could vaporize or pulverize a commercial or military reactor core or release stored waste and distribute it over a wide area (Pittock et al. 1989). Nuclear reactors on ships are also possible targets, and each nuclear propulsion system is fueled with highly enriched uranium that is the equivalent of 10–20 15 kt bombs. Accidents at the nuclear reactors located at Chernobyl, Ukraine, and at Fukushima, Japan, show that even if a bomb does not damage a reactor, they can be destroyed by their fuel overheating if cooling water is lost, which might occur if power and fuel for generators were lost for an extended period. Starr (2023) argues that EMP could be capable of destroying nuclear reactor cooling systems. Likewise, waste storage is vulnerable to being broken open or catching fire if cooling water is lost (Alvarez et al. 2003).

Nuclear power plants contain large amounts of radioactive materials not only in the reactor cores but especially in waste storage that may be near the reactor. It is estimated that there are 250,000 tons of highly radioactive nuclear waste stored worldwide, with almost 90,000 tons stored at 75 sites in the United States. Since the fuel rods in nuclear reactors are replaced about every 5 years, the waste material eventually contains more radioactive material than that in the reactor. The amount of radioactive waste is dominated by spent fuel and is estimated to be 28,000 EBq (IAEA 2008). The amount of radioactive material is measured with the unit of a becquerel (Bq), which is the amount of material that produces one nuclear decay per second. A becquerel is a very small amount of material, so it is often given a prefix that is the number of factors of ten multiplying the unit. The relevant unit for reactors and bombs is the ExaBq or EBq, which is 10^{18} Bq.

Two materials are of greatest concern. [131]Iodine is of concern because it is readily taken up by the body though its half-life is only 8 days, while [137]Cesium

Table 10.1 The Amount of Radiation Generated by Several Sources, in Ebq

Radiation Source	^{90}Sr	^{131}I	^{137}Cs
Fukushima	0.000035	0.1–0.5	0.006–0.2
Chernobyl	0.01	1.8	0.085
543 Nuclear weapons tests with 189 Mt fission yield	0.61	5.3	0.92

is of concern because of damage due to external exposure and it has a half-life of about 30 years. It is estimated that 50–60% of the ^{131}I and 20–40% of the ^{137}Cs in the core was released in the meltdown at Chernobyl (NEA 2002). In the case of Fukushima, it is estimated that about 1% of the ^{131}I and 1% of the ^{137}Cs in the core were released from the three reactors that were destroyed (Tanaka and Kado 2015). Chernobyl may have released a higher fraction of the radiation in its core than Fukushima because it had a steam explosion of the melted fuel that ruptured the containment vessel.

The radiation release from the Chernobyl reactor accident, the Fukushima reactor accident, and nuclear weapons tests are compared in Table 10.1. The specific isotopes released from a reactor are different from a nuclear explosion, but some of them are in common. The Chernobyl nuclear reactor accident released a similar amount of ^{131}I, about 10% as much ^{137}Cs, and about 6% as much ^{90}Strontium as all the above-ground nuclear tests. A Chernobyl-scale accident releases about 40 times the amount of ^{137}Cs as a 1 Mt weapon. There are about 440 operable nuclear reactors in the world. There are 93 in the United States located at 55 nuclear power plants. France has 56 operable nuclear reactors, China has 54, Russia has 27, Japan has 33, South Korea has 25, India has 22, Ukraine has 15, and the United Kingdom has nine. Therefore, there are potential nuclear power plant targets scattered across the world. If the 645 Mt of deployed nuclear weapons were all detonated in a nuclear conflict, and the 344 reactors in the countries just listed were destroyed in the war, then about 20 times as much ^{137}Cs could be released from the reactors as from the bombs. If the spent fuel, mostly in storage near the reactors, were also destroyed, potentially a factor of about 100 times that much radiation could be released as from the above-ground nuclear tests.

It is not legal to attack nuclear reactors unless they are being used to support the military. There have been many attacks on nuclear power plants under construction such as in Iraq and Iran, as well as attacks by revolutionary groups on nuclear facilities that had nuclear material, threats of attacks by airplane hijackers, actual attacks by deranged individuals, and security breaches by nuclear activists.

It is not clear whether nuclear reactors would be targeted in a nuclear conflict. For instance, it would not be in Russia's interest to target European reactors since Russia is downwind of Europe. In 2023 and 2024, Russia is occupying reactors in Ukraine, and some have been damaged by nearby explosions. Russia or the United States might choose to target reactors in each other's country as a means of contaminating large parts of the country. Even if reactors were not directly targeted, they might be damaged because they are close to other potential targets in a nuclear war. In addition, nuclear reactors that are shut down often need to use backup generators to provide power to cooling water to keep their cores from melting. The failure of the backup generators is what caused the accident at Fukushima. In a war, the backup generators are likely to run out of fuel, and the personnel needed to operate the reactors may have either been killed in attacks or starved to death in the nuclear winter described in later chapters of this book.

10.7 Estimated Casualties and Fatalities from a Global Nuclear War

Prior to the 1990s, a number of evaluations of the total deaths and injuries were made for global nuclear wars. These included deaths due to direct effects as well as deaths due to radioactive fallout from surface bursts. Some of these studies are summarized in Table 10.2.

The first notable aspect of Table 10.2 is that the number of weapons used in these studies prior to 1990 is much higher than the numbers now available as given in Table 7.2, even though none of the studies in Table 10.2 used more than about half of the deployed strategic weapons in their analyses. Most studies prior to 1990 included weapons with yields that are much larger than the yields of current weapons, which are almost entirely less than 1 Mt and average slightly more than 200 kt, as shown in Table 7.2.

The Middleton (1982) study found the highest casualties (fatalities + injuries) in Table 10.2, about a billion (10^9) people. This study assumed that major cities worldwide (except most of Africa, South America, and New Zealand) would be targets. Such worldwide attacks are not plausible since cities outside the combatant countries would not be likely targets. The Middleton study suggests about half of the population of the combatant countries, or half the world's urban population, would be killed in a nuclear war using about 25% of the strategic deployed arsenal at that time.

Table 10.2 Studies Prior to 1990 of Fatalities and Injuries in Nuclear War

Study Name	Number of Weapons	Total Yield (Mt)	Warhead Sizes (Mt)	Direct Fatalities/ Injuries
OTA (1979) *Counterforce attack on silos, bomber bases, and submarine bases*	9 ICBM fields + 49 bomber and sub bases			2–20 million US 10 million USSR
OTA (1979) *Military and economic targets*	8,985	7,800	0.1–20	U.S. 35-77% of population (70–160 million) range due to sheltering USSR 20–40% (64–100 million)
Middleton (1982)	14,747 (173 on Southern Hemisphere, 1,941 on cities, 701 on industry)	5,569	0.1–10	750 million/340 million
World Health Assembly (1987)*Counterforce*	2,839 on US 4,108 on USSR	1,342 on US 844 on USSR		25-64 million/23-35 million The higher values include fires.
World Health Assembly (1987)*Urban*	100 on US 100 on USSR	100 on U.S. 100 on U.S.S.R.	1	66 million/5 million US 77 million/16 million USSR
Harwell and Hutchinson (1989) *Urban and Counterforce*	12,600	6,000 (2,250 on military-surface, 2,250 on military-air; 1,250 urban-air, 250 urban-surface)		Hundreds of millions from direct effects in Northern Hemisphere

The World Health Organization (WHO; World Health Assembly 40 1987) and the Office of Technology Assessment (OTA 1979) each considered a variety of scenarios. Both studies included very specific "counterforce" wars in which only military targets such as missile silos, bomber bases, and submarine bases were targets. OTA found about 30 million fatalities, while WHO found 25–64 million would die out of 566 million people in the United States and the USSR. Most of the fatalities were caused by radioactive fallout on downwind urban areas. OTA also considered an attack on military and economic targets, which are often collocated with urban areas. They estimate 35–77% of the US population and 20–40% of the USSR population would die in such a war. The total deaths might be as high as 260 million. The USSR losses are lower than the US casualties due to their more rural population distribution and more centralized economic targets. WHO also considered a war in which 100 weapons, each with 1 Mt yield, were used against 100 cities in the United States and in Russia. About 28% of the US and USSR populations were killed using what was about 1% of the strategic deployed weapons available at that time.

Table 10.3 lists studies since 2000 of casualties when the numbers of weapons and their yields were reduced from the 1980s among the Cold War adversaries. Three studies have been published for a war between India and Pakistan. In the case of conflicts between India and Pakistan, there has been an upward progression of casualty estimates as the arsenals of India and Pakistan have grown. All three studies have focused on attacks on cities since that is where the greatest populations and where the greatest fatalities would occur. Toon et al. (2007) assumed five times as many weapons were used as did McKinzie et al. and estimated slightly less than five times as many fatalities. Toon et al. (2019) revisited this issue using estimates of the numbers of weapons in India and Pakistan by the end of 2020s, about five times as many weapons as in our 2007 study, and again found slightly fewer than five times as many fatalities. The fatalities grow slightly less than linearly with the numbers of weapons because less-dense urban areas are attacked as more or larger weapons are used. The yields of Indian and Pakistani weapons could grow from Hiroshima-sized fission weapons with yields near 15 kt, like those that they tested in 1998, to thermonuclear weapons with yields of 100 kt or greater. Fatalities in India and Pakistan could reach 50–125 million people, with the range depending on weapon yield from 15 to 100 kt, from the 250 nuclear weapons used on cities in this hypothetical war between India and Pakistan. During WWII, it is estimated that about 50 million people were killed over 6 years, not including those who died from disease and starvation (Clodfelter 1992). A war between Pakistan and India with 15 kt weapons could lead to fatalities approximately equal to those worldwide in WWII, and a war with 100 kt weapons could directly kill about 2.5 times as many as died worldwide in WWII, and in this nuclear war the fatalities could occur in a single

Table 10.3 Studies Since 2000 of Fatalities and Injuries in a Nuclear War

Study Name	Number of Weapons Used	Total Yield(Mt)	Warhead Sizes (kt)	Direct Fatalities/ Serious Injuries (Millions)
India Pakistan War				
McKinzie et al. (2001a)	10 on Pakistani cities and 10 on Indian cities	0.3	15	2.9 /7.7
Toon et al. (2007)	50 each on the most populated locations in India and in Pakistan	1.5	15	India 12.4/6 Pakistan 9.2/3.8
Toon et al. (2019)	100 on Pakistan 150 on Indian population centers	3.75–25	15–100	50–125 depending on yield
Global Nuclear War				
Toon et al. (2008) *Countervalue*	4,000 (1,000 on US; 200 each on France, Germany, Japan, and UK; 1,100 each on China and Russia)	400	100	China 166/54 France 15/4 Germany 17/5 Japan 34/12 Russia 52/12 UK 18/5 US 58/22 Total 360/114 Russia 8–12
McKinzie et al. (2001b) *Counterforce*	1,300 on Russia			Russia 8–12
Helfand et al. (2000) *Counterforce*	2,000 on 1,249 separate US Targets; 25% malfunction	1,100	550	US 52±2/9±1
Helfand et al. (2000) *Countervalue*	500; 25% malfunction	275	550	97±3

week. The world's annual death rate from all causes is about 56 million people per year (United Nations 2017). Therefore, a war between India and Pakistan in our scenario with 15 kt weapons could kill the same number of people in a week as would die naturally worldwide in a year, effectively doubling the annual global death rate.

Few studies of the consequences of a global nuclear war have occurred since 2000, possibly because after the disintegration of the Soviet Union most people thought nuclear war was no longer likely. McKinzie et al. (2001b) considered a counterforce war between the United States and Russia with the constraint that strict measures would be taken to minimize collateral fatalities. Most of the deaths they considered were the result of radiation falling on regions distant from the blast. They estimated fatalities in Russia that are like those estimated by OTA (1979) for a counterforce war.

Helfand et al. (2000) found a counterforce attack on the United States by Russia using 2000 weapons whose yield is about twice as large as the average weapon in use today resulted in 50 million fatalities. Many weapons were used on the missile fields in Montana, North Dakota, and Colorado-Wyoming-Nebraska. Similar predictions of radiation fields shown in Figure 10.11 stretch across much of the United States. However, only about 10% of the computed fatalities in this analysis resulted from the radiation, even in a set of extreme assumptions that the fission yield of the weapons was 80% and there was no sheltering from the radiation. Instead, most fatalities occurred from blast and fires in economic or political targets such as state capitals, centers of commerce, and power plants. In a second set of assumptions, Helfand et al. (2000) assumed that Russia attacks cities because military targets are guarded by a National Missile Defense system. Even though fewer weapons were used, the 500-weapon countervalue attack caused more fatalities, almost 100 million, than the counterforce war with 2000 weapons because more urban areas were attacked. In both scenarios, a large fraction of hospitals were destroyed and medical workers killed.

Toon et al. (2008) considered a countervalue war in which the United States and Russia had a nuclear war with 4,000 weapons with 100 kt yield aimed at population centers, including those in Europe, based on population as shown in Table 10.3. In the United States, with 1,000 weapons and 100 Mt of yield about 58 million fatalities occurred, while in Russia with 1,100 explosions and 110 Mt of yield, 52 million died. In total, the world war including NATO investigated by us in 2008 kills 360 million people, around seven times as many as died in the 6 years of World War II. Our 2008 estimates of fatalities from 4000 relatively low-yield weapons are about half those of Middleton (1982), who used almost 15,000 weapons to attack nearly every large city that might be a conceivable target worldwide. It is likely that many of the weapons in the Middleton study were used on targets with low populations, so that despite using four times as many weapons, only twice as many fatalities occurred, and of course world population is much greater now than in 1980.

These studies consistently show that attacks limited to strictly military targets, such as missile silos, lead to deaths in the millions to tens of millions mainly from radioactivity blowing into populated areas. However, as soon as economic and other targets that involve urban areas are considered, deaths rise into the hundreds of millions from blast waves, burns, and prompt radiation.

Chapter 11
Firestorms in Cities

11.1 Cities on Fire

A nuclear explosion is like bringing a piece of the Sun to Earth's surface for a few seconds (Figure 10.4). In addition to the tremendous burst of light and other forms of radiation, a supersonic blast wave follows. These short-lived hazards produce three immediate dangers to humans: prompt nuclear radiation, thermal radiation, and intense winds and strong pressure gradients in the blast wave. These dangers are considered to result in assured destruction because they are bound to occur, and we can calculate with some certainty the distance over which the destruction occurs, which varies greatly between the dangers and for different yields. There is a fourth danger, fires, whose areas are likely to exceed the entire area impacted by the other dangers. The heat from the explosion incinerates everything nearby, including people, animals, plants, and buildings. If you are unlucky enough to be close to ground zero, you would be immediately turned into cinders and sucked up into the mushroom cloud. Even far from ground zero fires can be started by the light pulse and by weak blast waves that provide fuel and ignition by breaking open buildings and cutting natural gas and power lines as well as exposing fires being used in the buildings such as the pilot lights in water heaters and stoves.

Everything burned within about a 13 km^2 (5 square miles) region roughly centered on ground zero at Hiroshima according to Glasstone and Dolan (1977). The Hiroshima fire was not the normal type of fire that we see in forests or even urban areas. Instead, it was a firestorm. Most normal large fires occur when there are high ambient winds due to a meteorological event. For example, in late December 2021, unusually strong winds due to an advancing weather system, with gusts exceeding 160 km per hour (100 mph), struck near Boulder, Colorado, where Brian lives. A fire broke out in a small shed in the adjacent city of Superior. The winds quickly spread the fires as burning embers blew downwind, skipping some neighborhoods and starting fires in others. Within a single day, more than 1000 houses burned in the cities of Superior and Louisville over an area of about 25 km^2 (10 square miles) in Colorado's most damaging fire ever. Remarkably, only two people are known to have died, even though thousands

Earth in Flames. Owen Brian Toon and Alan Robock, Oxford University Press. © Oxford University Press (2025).
DOI: 10.1093/9780197799734.003.0011

fled the fire with almost no warning. This type of wind-driven fire is often called a conflagration. Similar fires occur in forests.

The fire near Boulder was largely extinguished the next day due to a snowfall, but forest fires can burn for days or even weeks. Conflagrations and other forest fires usually burn along a narrow but long line. While the flames might extend perpendicular to the wind for 1 kilometer (0.6 miles) or more, the flame front is typically only tens or hundreds of meters wide. Such fires advance with the wind and propagate as the infrared light from the flames at the front irradiates fuel ahead of the front and starts the fuel burning. This method of fire propagation is similar to the way that the glowing skies from shooting stars originating from the Chicxulub impact crater started worldwide fires 66 million years ago and killed the dinosaurs. The spread of the Superior fire through an urban area is partly blamed on many of the houses being only 3 meters (10 ft) apart. Once one house was set on fire, the infrared radiation from it set adjacent houses on fire if they were close by. In very strong winds, flaming embers may be tossed far ahead of the fire front propagating the fire in highly flammable fuels such as grass, leaves, or small shrubs, while skipping over intermediate areas, as happened in Colorado in 2021.

Tokyo, Japan, experienced a conflagration in World War II on March 10, 1945, when hundreds of American bombers attacked it with incendiary bombs. The fire was spread by high ambient winds to cover about 41 km² (16 square miles), killing the largest number of people of any single bombing raid of the war and burning to death more people in 24 hours than ever before or since in the history of the world. Fortunately, many people were able to escape a painful death through incineration by running to escape the advancing flames.

A nuclear explosion would create conditions far worse than a normal forest fire or an urban fire such as occurred in Colorado because the fires would not be blown by the wind across an area from which people could flee ahead of the flames. Instead, the entire area would be set on fire nearly instantly by the light pulse. The blast wave that followed the flash of light could blow out the flames, but crumpled structures burn more easily, and fires would be reignited by burning embers, electrical sparks, and leaking gas pipes. Imagine how easily a house would burn with open gas lines or a filling station with gas pumps knocked over. Some people might still flee on the outskirts of the damaged area as the fires grew in size, but injuries from collapsing buildings and flying debris due to blast waves and from burns due to thermal and nuclear radiation would make escape very difficult even far from ground zero.

But what is potentially the most dangerous global impact of urban fires is the climate change they could produce from the large amounts of smoke. The amount of climate change depends on the amount of smoke that rises to high altitude.

11.2 Pyrocumulonimbus and the Rise of Smoke to High Altitude

In Hiroshima, the fires fully engulfed the city around ground zero within about 3 hours. Ambient winds were light. However, the heat from the burning city created a strong updraft, which caused strong winds to develop around the fire and rush in toward the fire center. This type of fire is called a firestorm. These winds rushed in and fanned the flames, spreading the fire within the area and creating a tall cloud containing smoke from the fire as well as radioactive debris. Of course, most of the cloud was made of water vapor from the air and from the burning fuels. Figure 10.6 shows a picture of the pyroCb cloud about 3 hours after the nuclear explosion. The top altitude of the cloud is not well known, but one study of the photograph by Baba et al. (2011) determined its height to be 16 km. The pyroCb at Hiroshima was considerably higher than the altitude of the nuclear mushroom cloud (Figure 10.5), which was estimated to be about 8 km. Indeed, the energy released in the fire at Hiroshima is estimated to have been more than 1,000 times the energy released by the bomb. The energy release in a fire is one factor controlling the altitude reached by the smoke cloud.

A pyroCb is a cumulonimbus cloud created by a fire. These clouds are common over intense forest fires, with 44–147 observed per year between 2013 and 2021 (Fromm et al. 2022). Cumulus clouds are convective clouds that grow into thunderstorms if the atmosphere is unstable enough and there is enough heating and water vapor in the lower atmosphere. A thunderstorm is a cumulonimbus cloud, with nimbus indicating that rainfall may occur. Very large thunderstorms often reach the tropopause, which is the boundary between two layers of the atmosphere, the troposphere where we live and the stratosphere. The troposphere contains about 90% of the mass of the atmosphere. In 2017, pyroCb clouds were observed to reach the tropopause in British Columbia (~12 km) above several intense forest fires, and in 2019/2020 there were pyroCbs that reached the tropopause (~16 km) in Australia above a series of intense forest fires. In both these modern cases, satellites observed that heating by the Sun lofted the smoke deep into the stratosphere, reaching altitudes of 20–30 km. That the smoke would rise has been predicted by climate models since the 1980s, but the fire in 2017 was the first to inject enough stratospheric smoke for the rise to be observed.

Numerical models have found that fires burning over a large area often produce smoke that rises to the tropopause. At midlatitudes, the tropopause is near 10–12 km altitude. Commercial airliners, which often breach the tropopause and fly in the lower stratosphere during their cruise, operate at similar altitudes. However, the tropical tropopause lies between 16 and 17 km, too high for passenger planes to reach. The height reached by smoke from fires increases in proportion to the area burning, the energy released, and the amount of water vapor in the air, but is decreased as ambient winds increase.

Following a 100 kt nuclear explosion, the diameter of the circle enclosing the area in which fires can be ignited by the thermal flash from the bomb is about 9 km (see Figure 10.1), about the same as the depth of the troposphere. Convective or cumulus clouds are often generated by warm buoyant bubbles of air near the surface. When the bubble is as far across as the troposphere is deep, the heated air will likely rise to the top of the troposphere. Of course, a hotter bubble, just like a hot air balloon with its burners lit, will rise higher than a cooler one. As the bubble rises it will expand and cool leading the water vapor in it to condense and form a cloud. The condensation of the water vapor will release latent heat that will warm the air, increasing its buoyancy. (The opposite process is familiar to us when water evaporates after a swim, cooling our body. The pulse of warmth one feels in a sauna when water is dumped on the hot coals is due to water condensing on our bodies and releasing latent heat.) The formation of the cloud thus drives the bubble even higher. Strong winds may push cold air over the fire or break the bubble into pieces, which will retard the lofting of the smoke.

Of course, clouds can rain and remove smoke. Limited observations of pyroCbs show that they do not produce very much rain and are not efficient at washing smoke out of the sky like normal clouds. The reason for this inefficient removal is that the smoke "overseeds" the clouds. Every cloud droplet in the sky forms on a preexisting small particle. Normally the concentration of these small particles is modest. Over the oceans there are around 100 particles per cubic centimeter, coming from sea spray, sulfur gases produced by the ocean biota, and long-range transport of pollution from the land. Over the land, particle concentrations may reach hundreds or thousands per cubic centimeter, coming from anthropogenic pollution, vegetation, dust storms, fires, and so forth. The smoke just above a fire produces much higher particle concentrations. With higher numbers of particles, there will be higher numbers of cloud droplets with smaller average sizes than in a cloud that did not form above a fire. Rain has difficulty forming when the drops have small sizes. We estimate from limited data from high-altitude samples of clouds above fires that about 20% of the smoke is removed before reaching its peak altitude. The complexity of determining the altitude reached by the smoke from fires has led to debate about the fraction of the smoke that will reach the stratosphere. The bulk of the evidence from the fire at Hiroshima, large forest fires, and numerical models suggests that nuclear firestorms will generally loft smoke within a few kilometers of the tropopause or above it.

It matters that smoke reaches the tropopause where sunlight can loft it into the stratosphere because it never rains in the stratosphere. Smoke from fires in the troposphere lingers a few days, or weeks at most, before it is removed by rainfall. Smoke in the stratosphere lingers for years before it slowly falls below the tropopause or is carried downward into the lower atmosphere by winds. The climate responds to loss of sunlight due to long-lived stratospheric smoke to a much greater extent than to low-altitude, short-lived smoke from a small forest fire. Each year forest fires and other types of fires around the world are estimated to inject about 85 million tons of smoke and 4.9 million tons of black carbon into the lower atmosphere (Andreae 2019). The smoke from forest fires does contribute to Earth's current climate, producing a small cooling overall because black carbon is only a small fraction of the smoke. Fortunately, the lifetime of this smoke is only days since it does not generally enter the stratosphere. Otherwise, significant global cooling would result, as discussed below.

11.3 Fuel for Fires

One of the most common questions we are asked about the climate change caused by nuclear weapons is why didn't they occur after the 543 above-ground nuclear tests conducted before the Nuclear Test Ban Treaty of 1963? The answer is very simple. The scientists knew very well that exploding nuclear weapons would start fires if fuel were available. Therefore, tests were conducted in deserts or on islands where there was little to burn. Following Iraq's invasion of Kuwait, many oil wells in Kuwait were set on fire. Some people, including Carl Sagan, speculated in advance that these fires might cause a large climate change. That did not happen for several reasons. The fires mainly burned at the end of February and during March 1991. The atmosphere in wintertime is relatively stable, suppressing smoke lofting. Also, the oil wells were small, isolated heat sources, not wide-area fires, so their smoke did not rise into the upper atmosphere. The fires were also extinguished fairly quickly.

Uncertainty about the availability of fuel and the ability to ignite it is a major factor in the military's refusal to account for the damage caused by fires from nuclear explosions. Most important targets are not located in deserts or on sparsely vegetated islands. Many military targets are in grasslands, such as in Colorado, Wyoming, Montana, and North Dakota where US missile silos are located. Many Russian intercontinental ballistic missiles (ICBMs) are hidden in forests. We can determine the amount of fuel if we know the fuel per unit area and the area that is burned. Typical tropical forests contain about 2 g/cm^2 of dry biomass, while temperate and boreal forests have about 0.9 g/cm^2 (Houghton 2005). Grasslands and non-forested regions with sagebrush and chaparral have a wide range of biomass, from around 0.05 to 2 g/cm^2 (Sikkink et al. 2009). These are relatively small amounts of fuel, and in addition fires in these types of fuels do not produce a lot of black carbon or soot. Therefore, targets in rural areas are usually not considered to be important for causing climate change following a nuclear war.

Many military bases and military-related industries are located in or near cities. Cities generally have much higher fuel loads than forests or grasslands. The fuel load is the mass of combustible material that might burn in a fire. The trunks of trees in forests, for example, often do not burn in forest fires. Tree trunks and other biomass fuels that are more than about 8 cm (3 inches) thick are hard to ignite, so they do not contribute to fire spread, which happens quickly. These are called dense fuels. They may burn eventually if the fire is not extinguished. Urban fuels are much different from forest and grassland fuels. Lumber usually dominates the fuels in urban areas. It is thought that in the United States, Russia, and Europe about 80% of the urban fuel is in the form of wood and lumber. Lumber is usually dry, unlike forest or grassland fuels that contain water that must be evaporated for a fire to burn. Therefore, lumber burns more readily than forest wood. Plastics, petroleum and other hydrocarbons, paper, cloth, and asphalt are widespread in urban areas. Most of these materials are highly flammable. According to Underwriters Laboratory, in 1990 you had 14–17 minutes to escape a house fire, but in 2021 people had only 2–3 minutes to flee from their burning houses and escape the heat and poisonous fumes. The reduced time to escape is due to the abundant synthetic materials in a modern house, mostly made of plastics and petroleum-based materials, and modern open floor plans, which promote exposure of the fuel to oxygen and ignitions. When your fire alarm goes off, don't wait. Run for the exit, or crawl if there is already smoke!

The smoke that comes from burning cities poses a threat to the global environment. While lumber dominates the fuel load, 60% of the smoke comes from the materials other than lumber. Modern materials, such as plastics, not only produce smoke with much more soot than lumber but also produce high levels of air pollution, especially toxic chemicals (such as dioxins, furans, and PCBs) from burning plastics and petrochemical derived materials.

There are few cities for which the fuel loads have been determined. Even a 200 kt weapon can ignite fires over an area of about 100 km² (40 square miles), as shown in Figure 10.1. Over such a large area the fuel load is likely to vary substantially between urban centers and suburbs. Ideally one could use a high-resolution database to account for all the variations. However, to date such calculations have not been possible because the fuels in various types of buildings are not well known. There was one large study in 1998 of fuel loads in San Jose, California, which was mainly suburban with single-family houses. The average fuel loading was estimated to be about 1.34 g/cm² (Simonett et al. 1998). Bush et al. (1991) found that for six moderate size American cities such as Nashville, Tennessee, the average fuel loading was 1.4 to 2.1 g/cm². Larson and Small (1982) suggested that within the inner 2 km radius of urban cores in three model cities circa 1980 fuel loadings were 23, 41, and 63 g/cm^2. The dense city center of Hamburg, Germany, was studied in detail after the fire-bombing and firestorm there during World War II. Various estimates range from 12 to 47 g/cm^2. Toon et al. (2008) showed that these data on fuel loads can be combined with population density data into a simple equation for fuel load per person. Each person in the developed world has about 11 metric tons of fuel in their houses, businesses, schools, shopping areas, and industrial areas. There are population databases for the entire world at a resolution of about 1 km^2, allowing fuel loads to be approximated across cities by using the fuel per person. Of course, the concept of fuel per person is an average over a wide range of habitats and is likely only appropriate when averaged over much of a city.

We have compiled estimates of the soot, the black carbon component of smoke, reaching within a few kilometers of the tropopause for several scenarios for nuclear conflict, for natural fires, and for the global fires caused by the dinosaur-killing asteroid. These are summarized in Table 11.1. The soot injections range over about 6 orders of magnitude, or about a factor of 1 million, between forest fires in 2017 that were observed to put smoke into the stratosphere, and the global fires after the asteroid impact that killed the dinosaurs for which the smoke is observed in the boundary rock left from the event. The details behind these estimates are summarized by the papers referenced in Table 11.1.

In Table 11.1, we see that the K-Pg fires put about twenty times as much smoke into the stratosphere per area burned as did the forest fires in 2017 and 2019–2020. There are probably two reasons for the greater efficiency of creating soot in the K-Pg case. First, the Earth was warmer in the Cretaceous than now, and CO_2 was higher, which may have made the amount of fuel per area burned higher than now since vegetation is thought to have been lusher. Perhaps this amounts to a factor of two. Second, as we discussed previously, current forest fires are often driven by high winds and are therefore ventilated by oxygen coming from the sides of the fires. So, the fires are oxygen rich. The K-Pg fires essentially burned the biota of the entire surface of Earth at once. In this case, no oxygen could be supplied from the sides of the fires, but only from aloft. As a result, the fires were likely oxygen starved. Laboratory data suggest that about ten times more soot is produced in oxygen-starved fires than in oxygen-rich fires. While the fires were oxygen starved near the ground where the fires burned, the K-Pg fires could not have consumed a significant amount of the Earth's oxygen. The biomass burned was likely about 2–4 g/cm^2. However, the mass of oxygen above Earth's surface is about 200 g/cm^2. Therefore, globally less than a few percent of the total oxygen in the atmosphere would be needed to burn the global biomass.

Table 11.1 The Estimated Amount (Tg, or millions of metric tons) of Soot Lofted into the Upper Troposphere or Stratosphere by Large Fires

Scenario	Mass of Soot (Tg)	Area Burned (km^2)	Soot/Area (g/cm^2)	Reference for More Information
2017 British Columbia wildfires	0.006	750–2,250	0.0003–0.0008	Yu et al. (2019)
2019–2020 Australian wildfires	0.023	5,300	0.0004	Yu et al. (2021)
50 urban targets attacked in India and 50 attacked in Pakistan with weapons of 15 kt yield	6.6	1,300	0.5	Toon et al. (2007)
150 urban targets attacked in India and 100 attacked in Pakistan with weapons of 15 kt yield	16.1	3,250	0.5	Toon et al. (2019)
150 urban targets attacked in India and 100 attacked in Pakistan with weapons of 50 kt yield	27.3	10,825	0.25	Toon et al. (2019)
150 urban targets attacked in India and 100 attacked in Pakistan with weapons of 100 kt yield	36.6	21,650	0.17	Toon et al. (2019)
250 urban targets attacked in India and 250 attacked in Pakistan with weapons of 100 kt yield	46.8	43,300	0.1	Toon et al. (2019)
1,100 urban targets attacked in China and in Russia, 1,000 attacked in the United States, and 200 attacked in France, in Germany, in India, in Japan, in Pakistan, and in the United Kingdom with weapons of 100 kt yield	180	381,040	0.47	Toon et al. (2008)
K-Pg asteroid impact	15,000	1.5x10^8	0.01	Wolbach et al. (1990) Bardeen et al. (2017)

The urban fires from nuclear conflicts in Table 11.1 produce amounts of soot per area burned that vary by about a factor of five primarily because as the yield of the weapons increases more and more of the target area consists of suburbs or wildlands as opposed to urban cores. Urban cores can have higher fuel loads than suburbs by factors of 50 or more. Currently, the yields of nuclear weapons are much smaller than they were in the 1980s because nuclear planners realized that lower-yield weapons do more damage per amount of nuclear fuel and so they reduced the average yield. We see in Table 11.1 that the greatest damage per area burned in an India-Pakistan war occurs for the smallest-yield weapons. Of course, some targets have more fuel also. We assume China and India have much more fuel in urban cores than Pakistan or the United States because Chinese and Indian cities have more people and higher population densities on average. This assumption is based on our finding in the developed world of the 1980s that each person had 11 tons of fuel associated with them. However, it is possible that many countries have less fuel per person because they don't use much lumber in their buildings or don't have the same contents in their houses. Global fuel loads are poorly known and an area in need of further research.

The urban fires in Table 11.1 produce about 1,000 times as much soot per unit area as the forest fires of 2017 and 2019–2020. This difference is the result of the factors we have just described. There is 10 to 50 times as much fuel in urban areas, the smoke production in oxygen-starved urban fires is 10 times higher than in an oxygen-rich forest fire, and the soot fraction of the smoke from urban materials is up 20 to 50 times higher than in forest fires.

11.4 Analogs to Nuclear Firestorms

During WWII, firestorms or conflagrations were produced by "conventional" bombing in Dresden, Hamburg, Darmstadt, and Tokyo, and by nuclear bombing in Hiroshima and Nagasaki. We know, therefore, that cities can burn, and if winds are light produce firestorms.

Kurt Vonnegut was a prisoner of war in Dresden during the fire there, and in his book *Slaughterhouse Five* (Vonnegut 1969) he described his experience:

> *He was down in the meat locker on the night that Dresden was destroyed. There were sounds like giant footsteps above. Those were sticks of high-explosive bombs. The giants walked and walked . . . So it goes.*
>
> *A guard would go to the head of the stairs every so often to see what it was like outside, then he would come down and whisper to the other guards. There was a fire-storm out there. Dresden was one big flame. The one flame ate everything that was organic, everything that would burn.*
>
> *It wasn't safe to come out of the shelter until noon the next day. When the Americans and their guards did come out, the sky was black with smoke. The sun was an angry little pinhead. Dresden was like the moon now, nothing but minerals. The stones were hot. Everybody else in the neighborhood was dead.*
>
> *So it goes.*

So even one day later, the sky was still so full of the smoke that it almost completely blocked the Sun, and this was from "conventional" bombing.

In 1906, San Francisco was ravaged when an earthquake broke water mains and started fires in the city. Figure 11.1 shows a painting of the fires and smoke (Hansen and Condon 1989). Jack London (1906), the famous author, was commissioned to describe what happened for a prominent magazine, *Collier's The National Weekly*, and reported from San Francisco Bay:

> *Within an hour after the earthquake shock the smoke of San Francisco's burning was a lurid tower visible a hundred miles away. And for three days and nights this lurid tower swayed in the sky, reddening the sun, darkening the day, and filling the land with smoke.... I watched the vast conflagration from out on the bay. It was dead calm. Not a flicker of wind stirred. Yet from every side wind was pouring in upon the doomed city. East, west, north, and south, strong winds were blowing upon the doomed city. The heated air rising made an enormous suck. Thus did the fire of itself build its own colossal chimney through the atmosphere. Day and night this dead calm continued, and yet, near the flames, the wind was often half a gale, so mighty was the force.*

This is a quintessential description of a firestorm, which can result from burning cities no matter how they are ignited. Figure 11.2 shows what San Francisco looked like after the fires, which is very similar to the appearance of Hiroshima in Figure 11.3. The damage in both cases is mostly due to the fires.

Figure 11.1 "Lurid flames sweep San Francisco in William Alexander Coulter's (1849–1936) panorama of the largest maritime rescue in United States history, where more than thirty-thousand people were taken from the shoreline between Fort Mason and the foot of Lombard Street" (Hansen and Condon 1989). Painting from http://www.sfmuseum.org/1906/coulter.html

Figure 11.2 This photograph, taken by George Lawrence from a series of kites five weeks after the great earthquake of April 18, 1906, shows the devastation brought on the city of San Francisco by the earthquake and subsequent fire. The view is looking over Nob Hill toward business district, south of the Slot, and the distant Mission. The Fairmont Hotel, far left, dwarfs the Call Building.

Photo courtesy of Harry Myers. http://earthquake.usgs.gov/regional/nca/1906/18april/images/sf06.city.html

To summarize, while there are numerous uncertainties about nuclear generated fires, nuclear explosions will almost certainly start fires in any area with exposed fuels because of the intense light flash caused by the explosion and because of shock waves creating ignition sources by breaking open buildings, gas lines, and electrical lines. Both Hiroshima and Nagasaki experienced large fires. However, we know little about how fires spread across cities, despite intense research on how fires spread in forests. Fuel amount is critical to producing a lot of smoke, but some fuel may be buried under fallen buildings and escape ignition. Fuel burial is likely important near ground zero where strong shock waves can knock down buildings, but the area with strong shock waves is likely only 10–15% of the area ignited. You can see that the rubble area is small by comparing the area covered by 10 psi shock waves, strong enough to knock down concrete buildings, with the fire ignition area in Figure 10.1.

The composition of the smoke is important, particularly the fraction of the smoke composed of soot, which is light-absorbing black carbon. Smoke from fires in vegetation is observed to contain only a few percent of black carbon. However, fuels such as oil, plastic, asphalt, and lumber produce smoke rich

Figure 11.3 A photograph of Hiroshima after the atomic bombing from the U.S. Air Force photo library, showing the city after some roads were cleared off. Almost all the buildings burned in the subsequent firestorm. For concrete buildings, many interiors burned out.

in black carbon, especially if the fuel is oxygen starved. Black carbon is what makes most smoke from urban fires look dark. The amount of smoke that may be removed by rainfall and the height reached by the smoke from urban fires are poorly constrained by observations. Observations of smoke rising high into the atmosphere show that fire-induced thunderstorms, pyroCb, are not very efficient at removing smoke by rainfall. The large amounts of smoke in the clouds produces small cloud particles and reduced rainfall. Recent observations of large forest fires do show that smoke can be emitted into the upper troposphere or lower stratosphere where it is subsequently lofted higher into the stratosphere by sunlight heating the black carbon in the smoke, as predicted by models of nuclear winter. This rise to high altitude is important because it never rains in the ambient stratosphere. Smoke in the lower atmosphere may linger for only a few days. However, smoke in the stratosphere may remain for years.

Living through a large urban fire is a terrifying experience as recounted in the stories of Kurt Vonnegut and Jack London. Imagine the terror caused by hundreds of burning cities. Most of the dinosaurs may have died in the fires produced by the asteroid impact in what is now the Yucatan Peninsula of Mexico. We know those fires were global from the soot left behind in the K-Pg layer. Even so, the fires did not directly cause the extinctions in the oceans; something else happened. Hundreds of burning cites would not cover a very large fraction of the Earth's surface area, so burning cities are not directly a threat to those far removed from the conflict. However, for both dinosaurs and people, stratospheric smoke from the fires can cause dramatic climate change after a nuclear conflict, threatening most of the life on the planet, as discussed in the next chapter.

Chapter 12
Climate Disaster, Climate Models, and Natural Analogs

12.1 Climate Disaster from a Global Smoke Cloud

The most dangerous global impact of nuclear fires is the climate change produced by the smoke. Today, nuclear weapons are much more powerful than those that incinerated Hiroshima and Nagasaki in 1945. If dropped on cities and industrial areas, in addition to the horrific direct effects previously described, they would produce prodigious amounts of smoke. Once the smoke reaches the upper troposphere, it would be heated by sunlight and rise into the next layer, the stratosphere. There are no clouds in stratosphere, and there is no rain. This means that the black smoke particles would continue to rise due to solar heating and would be blown around the world for years. The amount of climate change would depend on the amount of smoke, which we outlined in Table 11.1.

Before we describe in detail the potential impacts, we ask how we can know what would happen. Certainly, there has never been a large-scale nuclear war before, so we cannot actually observe what happened in the past. The smoke from the Hiroshima and Nagasaki fires, and the other 66 cities the United States burned in Japan during the summer of 1945 using conventional weapons, was not enough to detect a noticeable climate impact. Robock and Zambri (2018) estimated that 0.65 million tons (0.65 Tg) of smoke might have been produced by burning cities in Germany and Japan during 1944 and 1945 and calculated that Earth's surface temperature should have decreased by a couple of tenths of a Celsius degree (less than half a degree Fahrenheit), but the data available back then on the amount of smoke, the amount of solar dimming, and the surface climate were not good enough to detect significant changes. And even if they were, due to the chaotic nature of weather and climate variability, the small expected signal would have been within the natural variability of weather.

This lack of observables means we are left with two methods to figure out the climatic impact of nuclear war. One is climate modeling. The other is to look at analogs—that is, look at things that have happened and that tell part of the story. This includes past cities that have burned and produced firestorms, forest fires that have injected smoke up into the stratosphere, and the demise of the dinosaurs.

Earth in Flames. Owen Brian Toon and Alan Robock, Oxford University Press. © Oxford University Press (2025).
DOI: 10.1093/9780197799734.003.0012

12.2 Early Computer Models for Climate Change Due to Smoke

In 1982, the Swedish journal *Ambio* commissioned a special double issue on the environmental effects of nuclear war (AMBIO 1982). They composed a scenario for a large-scale nuclear war in which 5000 Mt of weapons (equivalent to 333,333 Hiroshima bombs) were dropped on targets in North America, Europe, and Asia, and invited different scientists to write papers on the effects on various parts of the environment. Paul Crutzen (winner of the Nobel Prize in Chemistry in 1995) and John Birks investigated the effects on atmospheric pollution (Crutzen and Birks 1982). One of the most important constituents of air pollution near Earth's surface is ozone, a major component of smog, which is formed by a photochemical reaction. Sunlight shining on a mixture of chemicals that would be released by explosions of industrial targets would produce ozone. To calculate the reaction rates, they needed to know the amount of sunlight. Following suggestions by Lewis (1979), they did a quick calculation of the amount of smoke that would be produced from forest fires that they thought would burn after the bombs exploded. They were amazed to find that there would be so much smoke that virtually no sunlight would reach the surface in the areas that were downwind of the places bombed! They also concluded that ozone would not be a major atmospheric pollutant near Earth's surface.

In parallel, in 1981, at one of the first meetings focused on the Alvarez et al. (1980) discovery, Brian presented simulations of the impact on climate of the asteroid impact that killed the dinosaurs. At that time, it was thought that the extinction event was due to the dust making up the thin geologic layer marking the extinction of the dinosaurs. An audience member suggested that Brian's group should look at similar climate effects from a nuclear war. Brian discussed this idea with Richard Turco, who happened to work at a company, R&D Associates, that did studies of nuclear weapons' impacts, and they began to investigate lofting of dust by nuclear explosions. However, they soon learned of the work by Crutzen and Birks (1982) that forest fires might be started by nuclear weapons. It had earlier been suggested by Lewis (1979) that urban fires would likely be more damaging than the blast waves considered by the military and that the smoke from urban and forest fires might impact the weather. Scientists at R&D Associates had been actively investigating urban fires set by nuclear weapons for many years. Turco computed the amounts of smoke generated in urban areas and found it was much higher than that from forest fires. Working with Tom Ackerman and Jim Pollack, they also found that the impacts of the smoke on the climate were much greater than from dust and, indeed, could cause temperature declines like those in the last ice age. Jim Pollack and Brian had been graduate students of Carl Sagan, and they invited him to join the group because they knew he was interested in nuclear wars, but for an entirely different reason.

Carl was surprised by the potentially immense numbers of intelligent civilizations in the Milky Way Galaxy computed from the straightforward equations devised by Frank Drake. Carl thought that something must limit the lifetime of intelligent civilizations; otherwise, we would have been visited by them. He thought nuclear conflicts were a plausible mechanism to destroy young civilizations. Eventually, this group became known as TTAPS, after it was pointed out in 1983 that the initials of the last names of the authors (Turco et al. 1983) spelled out that acronym.

The climate model used was very simple by today's standards. It treated the entire atmosphere of Earth as a vertical column of air in which detailed radiative transfer computations were done. While simple, the model had already been shown to replicate the effects of volcanic eruptions on climate and the possible extinction effects from an asteroid impact. In response to a NASA requirement that their work be reviewed, they wrote a draft paper describing this work called the Blue Book and organized a large meeting involving scientists from around the world to review it.

Simultaneously, an investigation of the Blue Book was initiated by the US National Academy of Sciences at the request of the Department of Defense. Climate researchers at the US National Center for Atmospheric Research (NCAR) quickly began studies with multidimensional climate models, as did others.

In the Soviet Union Vladimir Aleksandrov and Georgiy Stenchikov, who had received a copy of the Blue Book, published a three-dimensional simulation in 1983, as did Mike MacCracken (1983) in the United States. During 1984, there were fifty-six papers referencing the TTAPS study by scientists from several nations, including the United States, where Curt Covey et al. (1984) conducted three-dimensional climate simulations and Alan Robock (1984) used an energy-balance model. These researchers used many types of climate models to calculate how Earth's climate system would respond to so much smoke in the atmosphere and how the temperature, precipitation, and winds would change. The cooperative research between American and Russian scientists was a natural outgrowth of the close working relationship that had developed during the Soviet-American scientific exchange program in climate research from 1979–1995, formally known as Working Group VIII of the US–USSR Agreement on Cooperation in the Field of Environmental Protection.

Brian met Vladimir Aleksandrov, for example, prior to the Blue Book release while Vladimir spent time doing research on climate in the United States at the Lawrence Livermore National Laboratory and NCAR. Alan also met Vladimir at NCAR during a visit in the summer of 1978, when Vladimir spent 6 months there (Figure 12.1). In addition to US and Soviet collaborations on climate science, scientists from the US Department of Energy collaborated directly with university scientists and scientists from other government agencies in an

Figure 12.1 Vladimir Aleksandrov during a visit to the National Center for Atmospheric Research in Boulder, Colorado, on May 26, 1978.
Image from the University Corporation for Atmospheric Research, used with permission

unclassified, thorough investigation of nuclear winter. Finally, scientists representing the National Academies of many nations under the Scientific Committee on Problems of the Environment, a part of the International Council of Scientific Unions, investigated the physical impacts of nuclear explosions and the ecological and agricultural effects (Harwell and Hutchinson 1985; Pittock et al 1989) .

All these studies came to the same general conclusions. For a nuclear war between the United States, Europe, China, and the Soviet Union, there would be so much smoke that virtually all the sunlight would be absorbed, and the surface of Earth would become cold and dark for months, if not years. In the resulting nuclear winter, temperatures would plummet below freezing in the middle of continents, even in the summer, and agriculture would grind to a halt, sentencing most people on Earth to death by starvation.

While all these science studies took place, Carl engaged in a variety of communications efforts. On October 30, 1983, almost 2 months before the publication of the *Science* article by Turco et al. (1983), Carl published a popular description of the work in *Parade Magazine*, which was distributed along with Sunday newspapers to more than 10 million people. This prepublication release to the public is unusual in the scientific literature, and many critics used it to detract from the nuclear winter research, even though it is irrelevant to the science. Sagan also debated Edward Teller, so-called father of the hydrogen bomb, before Congress in 1984, leading to an ongoing debate between them. Of course, much of the nuclear establishment and especially the parts involving the Strategic Defense Initiative opposed the idea of nuclear winter, which threatened the entire concept of mutual assured destruction. Carl's heavy involvement with the press and high-level political figures led to him being targeted by critics of nuclear winter. Most of these attacks were personal rather than related to the science.

In science, every new theory or idea of importance becomes a target for research and analysis by others. It is normal for the science community to try to tear apart every aspect of a problem looking for errors or discrepancies. A similar flood of criticism and research into alternatives was aimed at the Alvarez discovery that an impact killed the dinosaurs, and debate about the science is ongoing to this day. When a theory survives such investigations over a long time period, it gives the theory credibility. That's as close as you can get to "truth" in science; tomorrow someone may yet come up with a counter to any theory.

However, Carl's fame and position in the public's mind as the leading scientist of the day led to unusual personal attacks. These were not all random; many were organized. Many of the same people who criticized Carl, and to some extent the rest of the TTAPS group, were the same figures involved as naysayers on cigarettes being bad for your health, that humans were causing the ozone hole by releasing chlorine compounds to the atmosphere, or that global warming is caused by fossil fuels. Brian recalls a well-known NASA scientist and space-exploration figure telling him that "working with Carl is like getting in bed with an elephant. If he rolls over, you are dead." Oreskes and Conway (2010) investigated some of these interactions in their book *The Merchants of Doubt*. While the Merchants created an atmosphere in which science is now disregarded by some modern politicians who find facts inconvenient, in the 1980s politicians sought the advice of the science community and made decisions based on careful studies by well-informed people.

In 1986, President Ronald Reagan of the United States and President Mikhail Gorbachev of the Soviet Union took the first steps in history to reduce the numbers of nuclear weapons. Every American president since then and every Russian leader since then through 2016 further reduced the number of these weapons, as we have illustrated in Figure 7.1. At the time, President Reagan and President Mikhail Gorbachev explained, as we quoted in Chapter 7, that the advice of scientists convinced them to reduce nuclear weapons. Unfortunately, the decline of nuclear weapons ended in 2016 with the Trump and Biden administrations, except for the backlog of weapons being dismantled.

In 1991, the Cold War ended, as did the Soviet Union, and so it seemed that the threat of nuclear war had also ended and world peace had been achieved. Turco et al. (1990) reviewed the status of the research into nuclear winter as fears of nuclear war diminished.

12.3 Modern Computer Models of the Environmental Impacts of Nuclear War

With the end of the Soviet Union and the Cold War in 1991, research into the effects of nuclear explosions paused, only to be reawakened a decade later by nuclear saber rattling by India and Pakistan. In 2001–2002, India and Pakistan had a military standoff with troops massing along the Line of Control in Kashmir in response to a terrorist attack on the Indian Parliament in New Delhi. This was their second standoff since both nations successfully tested nuclear weapons in 1998. A reporter called Brian to learn the consequences of a nuclear conflict between the two countries. As is typical, reporters are on a tight schedule and don't have time to wait for research to be done on their inquiries. Brian responded with a comment to the effect that many would die in India and Pakistan even from a single nuclear explosion, but the impact would not be felt very far beyond their borders. Brian felt guilty about this quick response, and the exchange with the reporter inspired him to investigate the nuclear arsenals of India and Pakistan. It took several years for Brian to find the time for this study, but he discovered that by 2005 India and Pakistan might each have 100 nuclear weapons with a yield like that of the bomb dropped on Hiroshima during World War II.

Estimates of the soot that would be produced in a war involving warheads targeted on cities led him to contact Rich Turco, and they devised a simple method to compute smoke using population databases that computer wizard Charles Bardeen, then a graduate student working with Brian, put into action. To their surprise, the estimated amount of smoke was very large, about 6 million tons (Table 11.1). Simple back-of-the-envelope climate calculations indicated the smoke would cause a significant climate change. However, Rich and Brian felt a simple climate calculation was not adequate to publish the results. Brian was familiar with Alan's recent work on volcanic effects on climate and his prior interest in nuclear winter. Brian and Rich approached Alan at the 2005 American Geophysical Union meeting in San Francisco, which was attended by about 10,000 geoscientists. Standing in a crowded hallway with people rushing past to attend exciting talks, they asked Alan if he would be interested in modeling the climate effects from the smoke they estimated from the putative India and Pakistan conflict. Alan was excited to learn about the result and realized Brian and Rich needed help from a climate modeler and that his graduate student Luke Oman, who was studying the effects of volcanic eruptions, was just the person to help. Georgiy Stenchikov, who had done the first three-dimensional climate model simulation of nuclear winter, was also working at Rutgers and became part of the team.

Modern climate models incorporate a complete treatment of the ocean and simulate the atmosphere from the ground through the mesosphere up to 80 km (50 miles) and beyond. They are able to simulate the climate for decades or even centuries and can compute the radiative properties of complex aerosols such as soot. It was known from climate models of the 1980s that smoke injected into the upper troposphere would rise into the lower stratosphere due to sunlight heating the smoky air and making it buoyant. However, those climate models were handicapped by the primitive computers of the day. The models only covered the atmosphere to near 20 km, and they could only simulate climate perturbations for a few months. In 2007, Alan and his collaborators (Robock et al. 2007a) used a modern climate model from NASA and found that the smoke from a war between India and Pakistan would rapidly rise to heights approaching the mesopause. The great height of the smoke, keeping it away from rain in the lower atmosphere, prolonged the cooling at the surface for many years and allowed the smoke to blow around the entire Earth, producing significant global effects. Mike Mills et al. (2008, 2014), repeated Alan's calculations using the NCAR climate model and found similar climate changes, but also revealed that the ozone layer would collapse due to intense heating of the stratosphere by the soot absorbing sunlight. Two groups of scientists in Europe, using independent models, found similar results (Stenke et al. 2013, Pausata et al. 2016).

Alan quickly realized that the climate effects following a war between the United States, Russia, Europe, and China would last much longer than previously thought. He also wanted to settle the claims from the 1980s that led some to be skeptical of the nuclear winter results because of the limitations of the climate models used then. He and his colleagues (Robock et al. 2007b) used the NASA climate model to show that nuclear winter was even worse than had been predicted in the 1980s. Coupe et al. (2019) and Bardeen et al. (2021) repeated the calculation using an NCAR climate model that considered the atmosphere in more detail and included processes ignored in the previous simulations. They allowed smoke particles to grow and fall out of the atmosphere faster as well as ozone to react to the stratospheric heating. Amazingly, they found almost identical results, including details of the patterns of climate change in differ- ent locations and seasons. Although there are still uncertainties, the harder we study this, and the more detail we put into our models, the firmer the conclu- sion gets that a nuclear war between the United States and Russia, even using the current arsenals that are much smaller than at the height of the arms race in the 1980s, would produce nuclear winter. Also, conflicts between countries such as India and Pakistan, though not producing a nuclear winter in which tempera- tures remain below freezing at midlatitudes for a prolonged period of time, still cool the Earth enough to produce damaging global effects.

Figures 12.2–12.4 illustrate the range of effects that occur for different lev- els of soot injected into the upper troposphere for various scenarios of nuclear wars between India and Pakistan and for a global nuclear war. The smoke from the scenarios in Table 11.1 was rounded downward for the 5 Tg war with 100 weapons of 15 kt and for the 150 Tg global war. Figure 12.2 shows that even for a 5 Tg war with 100 weapons there is more than a 5% loss of sunlight in the first 2 years after the conflict, and losses continue for about 10 years. As the amount of soot produce by the urban fires increases so does the loss of sun- light, until more than 70% is lost in the first year following a global nuclear conflict.

With less sunlight available, the global surface temperature declines, as shown in Figure 12.3. Even a war producing 27 Tg of soot (250 urban areas attacked with 50 kt weapons in India and Pakistan) is followed by global average tem- peratures like those found in the last Ice Age, which are marked by the vertical bar. Of course, in an ice age the cold temperatures persist for tens of thousands of years, while for the nuclear conflict they persist for 5 years and return to nor- mal after about 15 years. The lowest temperatures occur about 3 years after the soot injections. Precipitation is controlled by evaporation of water mainly from the oceans. As sunlight is reduced, precipitation will decline worldwide. Figure 12.3 shows losses of precipitation from 10 to 60% depending on the number and yields of the weapons and their targets.

Figure 12.4 shows the changes in ocean surface and land surface temperatures instead of the global average temperatures shown in Figure 12.2. The land cools much more than the oceans. Only the upper few meters of ground are involved

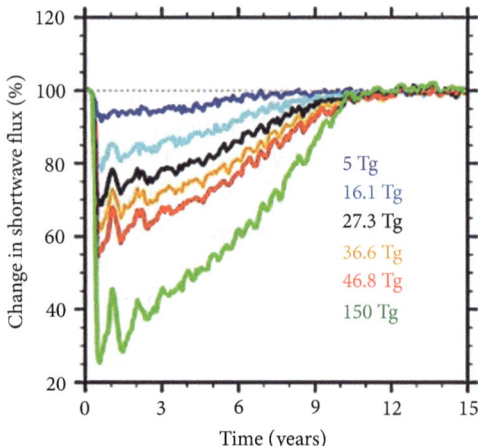

Figure 12.2 The percentage change in the solar energy reaching the Earth's surface for the various nuclear war cases listed in Table 11.1.
From Toon et al. (2019)

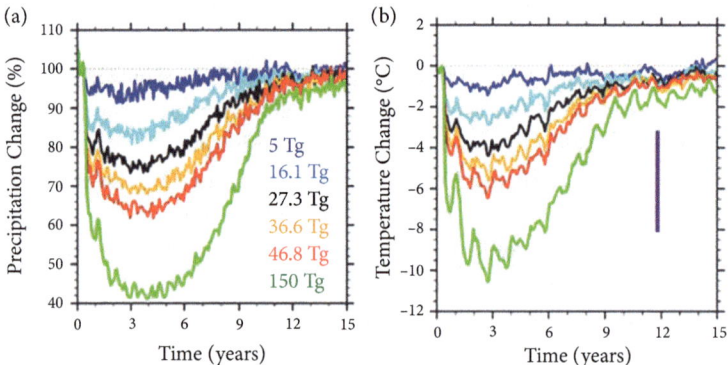

Figure 12.3 The percentage change in precipitation at the Earth's surface and the change in global average surface temperature for the various nuclear war cases listed in Table 11.1. The vertical bar denotes the estimated range of temperatures during the last Ice Age.
From Toon et al. (2019).

in the land surface temperature change, while hundreds of meters of the ocean are involved in changing its temperature since sunlight penetrates deeply into the oceans and cold ocean water sinks. The greater mass of cold water than cold land leads to the ocean cooling less than the land. The land also shows seasonal swings in temperature, which are muted in the oceans due to the larger mass involved.

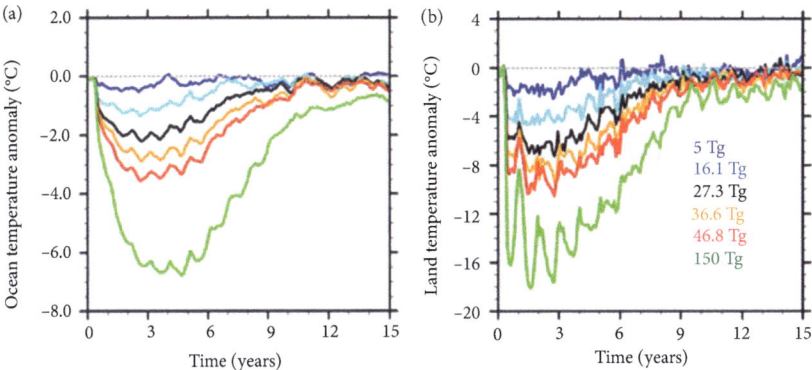

Figure 12.4 The change in ocean and land surface temperature for the various nuclear war cases listed in Table 11.1. Note different y-axis scales.
From Toon et al. (2019)

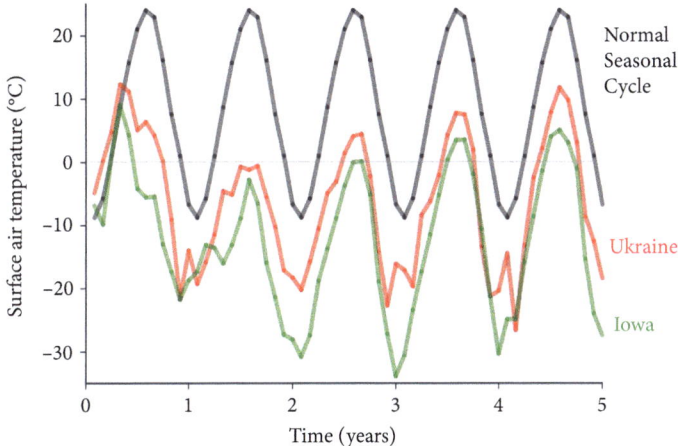

Figure 12.5 The daily minimum temperature in Iowa and in the Ukraine following a nuclear conflict in which 150 Tg of soot is placed into the upper troposphere. Also shown are typical temperatures for years without a war. Tick marks are on January 1 of each year.
Based on results in Coupe et al. (2019)

Figure 12.5 shows a detail of the temperatures for Iowa and the Ukraine following a nuclear war with 150 Tg of soot in the stratosphere. Iowa and the Ukraine are major agricultural areas for the United States and Europe. Following the start of the war in May, summertime daily minimum temperatures quickly fall below freezing. It is not until July of the second year after the war that either region experiences a single day without some subfreezing temperatures. Such persistent low temperatures define a nuclear winter. Agriculture is clearly not practical in such conditions.

12.4 Natural Analogs for Climate Changes Due to Smoke

How can we investigate the extreme situations that have been calculated in our theoretical models? We cannot bring the Earth into the laboratory and perform a nuclear war experiment on it. And we cannot actually perform the experiment of having a nuclear war in nature. Or we could only perform it once, and then it would be too late. We can, however, look back at other occurrences in the climate system to see if any similar situations have existed that would let us learn about what would happen in the event of a nuclear winter. We can search for situations that teach us about parts of the interactions discussed above, or for the global climate response, and we need not confine the search to our planet.

There are a variety of natural analogs for climate changes after a nuclear war, in which climate models simulate similar losses of sunlight. The most obvious are the day-night cycle and the summer-winter cycle. Imagine if the Sun did not rise tomorrow, if tonight was followed by another night, and then by another night. During the night, sunlight is reduced by 100% in most places because the Sun is on the opposite side of the Earth as during the day due to the Earth's rotation about its axis. As a consequence, nighttime temperatures drop rapidly, though the rate of decline depends on cloudiness, the local water vapor content of the air, and other factors. During the winter, the winter pole faces away from the Sun, and during the summer, the summer pole faces toward the Sun. These changes are due to the Earth's daily rotation axis not being perpendicular to the plane in which Earth orbits the Sun each year. The changes are not due to the Earth being farther from the Sun in winter. In fact, Earth is slightly farther from the Sun in July and slightly closer in January. Due to tilt of the spin axis, at a typical midlatitude location the average daily insolation in January is about 25% of its value in July. Winter in continental interiors at midlatitudes is usually not conducive to agriculture. These examples all illustrate that when sunlight is reduced rapid cooling occurs. Of course, the climate models used to compute the climate changes after nuclear conflicts duplicate these diurnal and seasonal changes in temperature, taking many important factors into account such as the altitude of the site and its distance from an ocean or sea. Figure 12.5 shows that for Ukraine and Iowa temperatures drop very rapidly following a nuclear conflict in May due to absorption of sunlight by the smoke in the stratosphere, despite there still being lots of sunlight entering the atmosphere since it is summer. There remains a seasonal cycle of temperatures with lowest values in winter, and the difference in minimum temperature between the warmest summer days and the coldest winter days is about the same for the climatology and the nuclear case.

Periodically, clouds of dust are blown up from the Sahara Desert and transported all the way across the Atlantic Ocean to the United States. From this, we learn that dust particles in the troposphere can be spread large distances around the world. It has been observed that under Saharan dust clouds it is colder and there are fewer water clouds and less rain. This suggests a similar reaction of climate to nuclear soot and dust.

When the US Mariner 9 spacecraft first flew by Mars in November 1971 to take high-resolution pictures of the surface, the entire Martian globe was covered by a thick cloud of dust. (Mars has an atmosphere, too, but much thinner than Earth's.) The heating of the atmosphere caused by the initial local dust storm induced a circulation that transported the dust upward and into the other hemisphere (Haberle et al. 1982). This same effect is part of the nuclear winter scenario and implies that regions far removed from the conflict would experience climate changes.

There have been large forest fires in recent history, too, although not as large as those implicated in the extinction of the dinosaurs or nuclear winter. We can study these to learn about the properties of the smoke particles and how they affect light and heat radiation going through them. In September 1950, a giant forest fire raged in western Canada for a week (Robock 1991). A week later, the smoke cloud covered the eastern United States, and a week after that it was seen over Europe. Again, we see how far particles can be transported by the wind before getting washed out of the atmosphere. When the smoke was over Washington, DC, weather forecasts for high temperature were as much as 6°C (10°F) too high. The actual temperatures were 6°C (10°F) lower than were forecast because the sunlight was blocked by the smoke. Even in late September, when the sunlight is not very intense, large surface temperature effects can result from a smoke cloud in the atmosphere, one that has much less opacity than that calculated for nuclear winter.

This anecdotal report for 1950 was confirmed in more recent studies that found surface cooling of 2–4°C (4–7°F) in the Midwestern United States under smoke clouds generated by forest fires in British Columbia, Canada, in 1982, and more than that produced by the September 1988 forest fires in Yellowstone Park (Robock 1991). Forest fires in northern China in 1987 and more recent ones in Australia and Canada produced thick smoke clouds that rose all the way into the stratosphere, where they resided for a long time, just as predicted by the computer models discussed in the previous section for other fires with stratospheric smoke. In a nuclear winter situation, the smoke from burning cities and industrial facilities is expected to be thicker and blacker than forest fire smoke, producing even more cooling.

In 2017 and again in 2019–2020, large smoke injections into the upper tro-posphere and lower stratosphere were observed. Western Canadian wildfires during August 2017 injected smoke into the stratosphere that was detectable by satellites for more than eight months (Yu et al. 2019). The smoke plume rose from 12 to 23 km (7 to 13 miles) within two months owing to solar heating of black carbon, extending the lifetime and latitudinal spread. And between December 29, 2019, and January 4, 2020, fires in southeastern Australia pro-duced at least 18 pyroCb clouds, pumping three times as much smoke as the Canadian fires into the lower stratosphere. Again, the Sun heated the smoke, lofting it from the lower stratosphere (15–16 km, 9–10 miles) to altitudes above 31 km (19 miles) in less than two months. The same models used for nuclear winter predictions duplicated the rise and dispersal of the smoke in detail (Yu et al. 2019). While changes in stratospheric temperatures and chemical compo-sition have been observed and simulated correctly by the models, they are not large enough to impact humans, because as shown in Table 11.1 the amount of smoke in the stratosphere was very small.

Another test of the model's ability to correctly calculate the climate response to particles in the stratosphere occurs for volcanic eruptions. Volcanic eruptions are quite common. There is a small one every year or two that places particles into the stratosphere. These cause observable changes in a variety of climate parameters, such a surface temperature, precipitation, solar energy reaching the surface, and stratospheric ozone. However, the impacts of these small eruptions, as well as those from forest fires, are too small to affect the average person.

Clouds of volcanic dust and sulfuric acid droplets that get injected into the stratosphere have been observed to be spread completely around the globe in three weeks and remain for several years. This is the same fate postulated for nuclear smoke that either gets initially injected into the stratosphere or is lofted there by solar heating. Large volcanic eruptions can also produce clouds in the troposphere immediately after the eruptions. Large surface local temperature changes have been observed under these tropospheric volcanic clouds follow-ing the 1883 Krakatau, 1980 Mt. St. Helens, and 1991 Pinatubo eruptions in a manner similar to the effects of forest fire smoke.

The long-lasting stratospheric clouds, composed of sulfuric acid, also have been observed to produce global climate changes for several years following large volcanic eruptions. One of the largest in recent memory was the eruption of Tambora in 1815, which was followed by such cold weather during the fol-lowing summer that 1816 has become known as the "year without a summer." That summer the famous poet Lord Byron lived by the shore of Lake Geneva, Switzerland, next door to his friends, Percy Bysshe Shelley and his young 18-year-old bride Mary. The weather was so cold and gloomy that they could not go boating or hiking, so they had a contest to see who could write the best ghost story, and Mary Shelley won by writing *Frankenstein*, which begins and ends with frigid images of the monster climbing over ice floes. Byron (1816) himself was so depressed by the cold, gray weather that he wrote a poem called *Darkness*, which begins

> *I had a dream, which was not all a dream.*
> *The bright sun was extinguish'd, and the stars*
> *Did wander darkling in the eternal space,*
> *Rayless, and pathless, and the icy earth*
> *Swung blind and blackening in the moonless air;*
> *Morn came and went–and came, and brought no day,*
> *And men forgot their passions in the dread*
> *Of this their desolation; and all hearts*
> *Were chill'd into a selfish prayer for light:*
> *And they did live by watchfires— and the thrones,*
> *The palaces of crowned kings— the huts,*
> *The habitations of all things which dwell,*
> *Were burnt for beacons; cities were consumed,*
> *And men were gather'd round their blazing homes*
> *To look once more into each other's face*

This remarkable description of what the world might be like in a nuclear winter was inspired by a volcanic cloud much thinner than a smoke cloud that would produce a nuclear winter.

Some volcanic eruptions have put massive amounts of particles into the stratosphere. President Ronald Reagan, commenting on why nuclear weapons should be eliminated, recognized that the 1815 eruption of Mt. Tambora caused cooling in some places that was so extreme that the year 1816 was called "the year without summer" or "1800 and froze to death." The 1991 eruption of Mt. Pinatubo in the Philippines was the largest of the twentieth century. It created about 30 Tg of stratospheric particles. This particle mass is similar to the masses of soot in Table 11.1 for some of scenarios for conflicts between India and Pakistan. You might wonder why 1992 didn't have climate changes like those outlined in Figures 12.2 and 12.3. First, the volcanic particles have a much shorter lifetime in the stratosphere, which limits the time that they can influence surface climate. After a year, only about a third of the particles remain, whereas for soot this takes five to seven years because it is lofted to very high altitudes by solar heating. Second, volcanic particles are largely made of sulfuric acid, which is a transparent liquid that looks like water. Hence the particles do not absorb much sunlight. Instead, they reflect some of the light back to space, which does act to cool the surface. In 1992, the loss of sunlight over Hawaii was about 2.7%

on average for 10 months after the eruption (Stenchikov et al. 1998, Dutton and Christy 1992), and there was a small observed global average cooling of a few tenths of a degree Celsius. However, the same mass of soot would produce a loss of solar radiation at the surface of 20–30% and a cooling of 3 to 4°C according to Figures 12.2 and 12.3. When light passed through the Pinatubo cloud, 20–30% of the light looking directly at the Sun would have been lost. A similar loss would occur for light passing through the same mass of soot particles. But the transparent volcanic particles also transmit a lot of the light downward as diffuse light, creating a bright glowing sky, especially at sunrise and sunset. Auto accidents were caused by the bright twilight skies after the Pinatubo eruption. As a result of the glowing skies, the amount of sunlight reaching the ground, which is the sum of that coming directly from the Sun and the amount coming from the sky, was little changed by the volcanic particles. In contrast, black soot absorbs most of the light and reflects only a little into the sky. Therefore, there is a large loss of light at the ground. The same models used for nuclear winter calculations have also been used for volcanic cloud calculations and are found to reproduce the properties of the particles, their impacts on sunlight and the temperature changes in agreement with observations.

Both climate models and analogs show that nuclear winter would result from a war between the United States and Russia, and even a war using many fewer nuclear weapons would be catastrophic.

12.5 Comparisons of Climate Changes after the K-Pg Event and Nuclear Wars

In Chapter 4, we examined the geologic observations of smoke at the K-Pg boundary produced by an asteroid impact coupled with extinctions both on land and in the oceans. As we pointed out in Chapter 4 and in Table 11.1, about 15,000 million tons (or 15,000 Tg) of smoke was placed into the atmosphere after shooting stars from the asteroid ignited the global forests and grasslands at the K-Pg boundary 66 million years ago. Figure 12.6 shows the changes in climate parameters for a variety of large soot injections ranging from 15,000 Tg to 1500 Tg, 150 Tg, and 16 Tg. The 15,000 Tg case represents the K-Pg extinction event. The 150 Tg case represents a global nuclear conflict. The 16 Tg case represents a war between India and Pakistan with 250 nuclear weapons each having a yield of 15 kt, like the yield of the nuclear weapon that destroyed Hiroshima during World War II.

As shown in Figure 12.6a, only the K-Pg event produces several years without enough sunlight to drive photosynthesis. This loss of light and stoppage of photosynthesis likely caused the observed extinctions in the oceans. The loss of

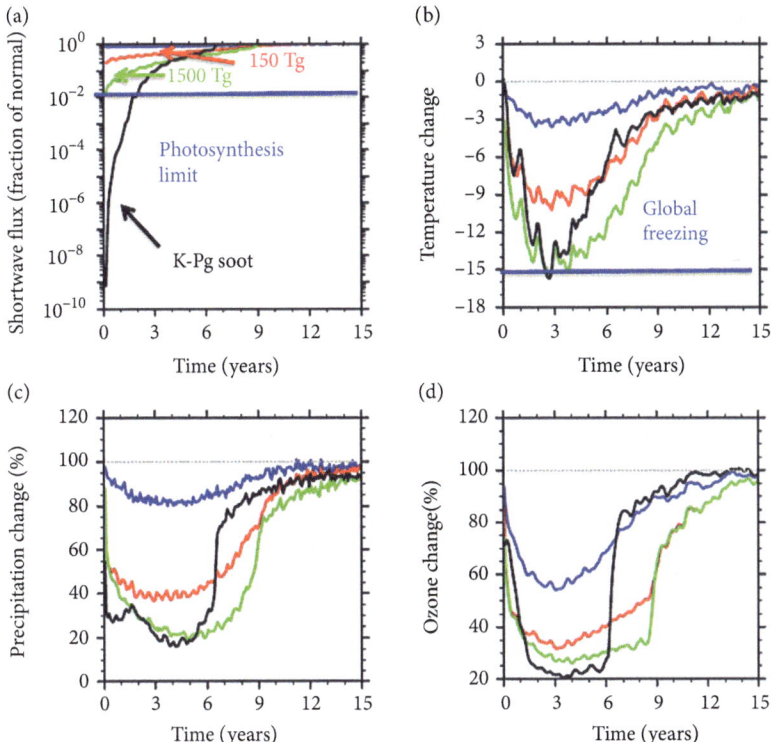

Figure 12.6 The change in sunlight, global surface temperature, precipitation, and ozone for the various cases listed in Table 11.1. Simulations are illustrated spanning four orders of magnitude of soot injection. The blue curves are for the 16 Tg war between India and Pakistan, the red curves are for the 150 Tg global war, the green curves are 1,500 Tg, and the black lines are for the K-Pg case of 15,000 Tg. Also shown in Panel (a) is the limiting amount of sunlight for photosynthesis, and in Panel (b) the temperature decline that represents a global average temperature at the freezing point.

Data from Toon et al. (2019), Bardeen et al. (2021), and Coupe et al. (2019)

light also led global surface temperatures to plunge for about 3 years and remain below normal for about 15 years. The global average surface temperature after the 15,000 Tg injection drops to the freezing point, probably dooming many creatures that survived the near-global fires that created the smoke. The global nuclear war has about a 70% loss of light for several years, as shown in more detail in Figure 12.2.

As shown in Figure 12.6b, the 150 Tg soot injection prolongs the cooling, which is near a global average temperature corresponding to freezing, longer than in the 15,000 Tg case, because the larger mass of the 15,000 Tg injection leads to bigger particles that fall out faster. The magnitude of the cooling is similar between the 15,000 Tg injection after the asteroid impact and the 150 Tg injection after a United States, Russia, Europe nuclear war.

All the soot injections lead to large losses of precipitation and ozone. As discussed in the next chapter, the increase in ultraviolet light at the surface, which damages DNA and is dangerous for life, is initially greatest for the smallest soot injection because the soot itself will shield the surface from the harmful rays. For the K-Pg case, as the smoke clears the surface is bathed in intense ultraviolet radiation.

Although we discuss in detail a scenario of a nuclear war between India and Pakistan, or between the United States, Russia, and their allies, the climate response and agricultural impacts depend only on the amount of smoke that is injected into the stratosphere. The multiyear lifetime of the smoke means that no matter where it is injected, it will spread around the entire Earth and be nearly equally spread between the hemispheres. In Chapter 8, we discuss many other scenarios of how a nuclear war might be started, resulting in large amounts of smoke that could be injected into the stratosphere. We only use the India-Pakistan case as one example, and do not mean to imply that they are the only ones who could produce this tragedy. Rather, our results imply that fires caused by nuclear weapons from any nation would contribute to a tragic nuclear holocaust.

All the changes illustrated in Figures 12.2–12.6 can trigger agricultural failures and expose life to harmful ultraviolet radiation, as discussed in the next chapter.

Chapter 13
Impacts on Humans of Nuclear War

13.1 Our Limited Food Storage

Regional crop failures occur almost every year, mainly due to bad weather. Occasionally these cause significant problems. For instance, in the 1970s the administration of President Carter shipped the Soviet Union the equivalent of 30% of the US wheat production to compensate for agricultural losses there. In 2010, due to bad weather, Russian wheat production fell 32.7%, Ukrainian wheat production fell 19.3%, Canadian wheat production fell 13.7%, and Australian wheat production fell 8.7% (Sternberg 2013). In the fall of 2010, China experienced drought and began to purchase wheat, which drove up prices. People in Middle Eastern and North African countries spend large fractions of their income on food. Libya, Jordan, Algeria, Tunisia, Yemen, and Egypt are each in the range of 35 to 44% of income spent on food. A large fraction of the food is bread. A significant causal factor in the Arab Spring uprisings, beginning in December 2010, was food insecurity due to the wheat failures in 2010 (Sternberg 2013). In the years following 2010, leaders were forced out of power in Tunisia, Egypt, Libya, and Yemen. Much of the turmoil in Syria between 2011 and the present started in response to drought-related crop losses in that country.

While these examples show that major problems can occur in regional food supply, generally losses in one area can be made up by transport of food from other regions that may be experiencing a bumper crop. This global-scale transport is at the heart of modern food supply. Most large cities have only about a week of food in storage and rely on constant transport. Unfortunately, the world is not prepared for a global food shortage, which might occur due to global climate changes after a large volcanic eruption, an asteroid impact, or a nuclear war.

Many are familiar with the story from the Bible or the Koran about Joseph and the pharaoh of Egypt. Joseph's interpretation of the pharaoh's dreams led the pharaoh to store grain from seven years of good harvests, which he then used to feed the Egyptians during seven years of bad harvests (Schneider and Mesirow 1976). However, the modern world does not have seven years' worth of grain in storage. Figure 13.1 shows that the yearly ending stocks of grain could feed the world from two to four months if no new food were produced. Thus,

Earth in Flames. Owen Brian Toon and Alan Robock, Oxford University Press. © Oxford University Press (2025).
DOI: 10.1093/9780197799734.003.0013

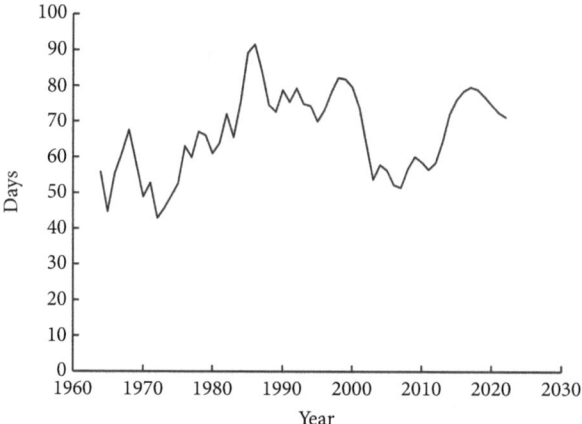

Figure 13.1 Number of days that the world could be fed from crops in storage. Data from Institute for Progress (ifp.org, downloaded May 16, 2024)

instead of seven years' worth of grain in storage, we have only a few months' supply of food in storage.

13.2 Analysis of Agricultural Losses after a Limited War

Agriculture is big business, and consequently agricultural models have been developed to predict how changes in temperature, rainfall, and other factors will impact food production. Our climate models, as well as natural analogs such as volcanic eruptions, tell us that a nuclear conflict will lead to cooler temperatures, lower precipitation, and diminished solar radiation. Agriculture is very sensitive to each of these climate variables. A variety of mechanistic agricultural models has been used to investigate the impacts from a range of smoke amounts. Mechanistic models forecast crop yield by simulating the underlying physiological processes in relation to the environment such as photosynthesis and respiration.

The smallest soot injections we have considered come from a conflict between India and Pakistan involving 100 weapons with 15 kt of yield used on cities in the two countries. This scenario was most relevant around 2007, when the arsenals of India and Pakistan were much smaller than now, but the scenario might also apply to a limited conflict or new nuclear states in the future. Using six global mechanistic crop models to look at the impacts of 5 Tg of stratospheric soot from a war between India and Pakistan in spring, Jägermeyr et al. (2020) found that the annual average global caloric production from the four major food crops— maize (corn), wheat, rice, and soybeans—would fall by 13%, 11%, 3%, and 17%

on average each year over five years. Total single-year losses of 12% would be four times larger than ever before since extensive records began to be kept in 1961 and would exceed impacts caused by historic droughts and volcanic eruptions. They found that domestic reserves and global trade could largely buffer the production anomaly in the first year. After the first year, current grain reserves would largely be depleted. Persistent multiyear losses would constrain domestic food availability and propagate to the Global South, especially to food-insecure countries. By year 5 after the nuclear war, maize and wheat availability would decrease by 13% globally and by more than 20% in 71 countries with a cumulative population of 1.3 billion people. It is assumed in this work that exports are banned once food reserves are depleted.

13.3 How Agricultural Losses Mount with Increasing Amount of Soot

Lili Xia et al. (2022) used the Community Earth System Model version 2 (CESM2) coupled with the Community Land Model version 5 crop model (CLM5crop) to compute the changes in maize, rice, soybeans, spring wheat, sugar cane, and cotton, as well as natural vegetation such as grasses, for nearly every country in the world and for the various scenarios outlined in Table 11.1 and Table 13.1. Figure 13.2 shows the changes in a variety of climate variables across all crops studied.

India and Pakistan are rapidly increasing the number of weapons in their nuclear arsenals and claim to be increasing yields over those tested in 1998. Toon et al. (2019) estimate that a nuclear war between them in 2025, in the most pessimistic case we investigated, could generate almost ten times the amount of smoke as our calculations for the 2005 arsenals, whose agricultural impacts were investigated by Jägermeyr et al. The aqua line in Figure 13.2a shows that for 47 Tg of smoke, the cropland average cooling would be five times larger than the 5 Tg cooling. Because the additional smoke is partially obscured by the other smoke above it, the increased temperature response is only half as large as the smoke increase. For a full nuclear winter with 150 Tg of smoke, which is three times more smoke than the 50 Tg case, there is only about one and a half times more cooling. This means that a nuclear war between two new nuclear states, India and Pakistan, using less than 2% of the global nuclear arsenal generating 47 Tg of soot, could produce two-thirds of the temperature change as the nuclear winter following a war between Russia and NATO.

Figure 13.2c shows that as smoke levels increase less sunlight is available at the surface. Sunlight is the fundamental driver for the other changes. Cropland precipitation (Figure 13.2e) falls significantly from the global average value of

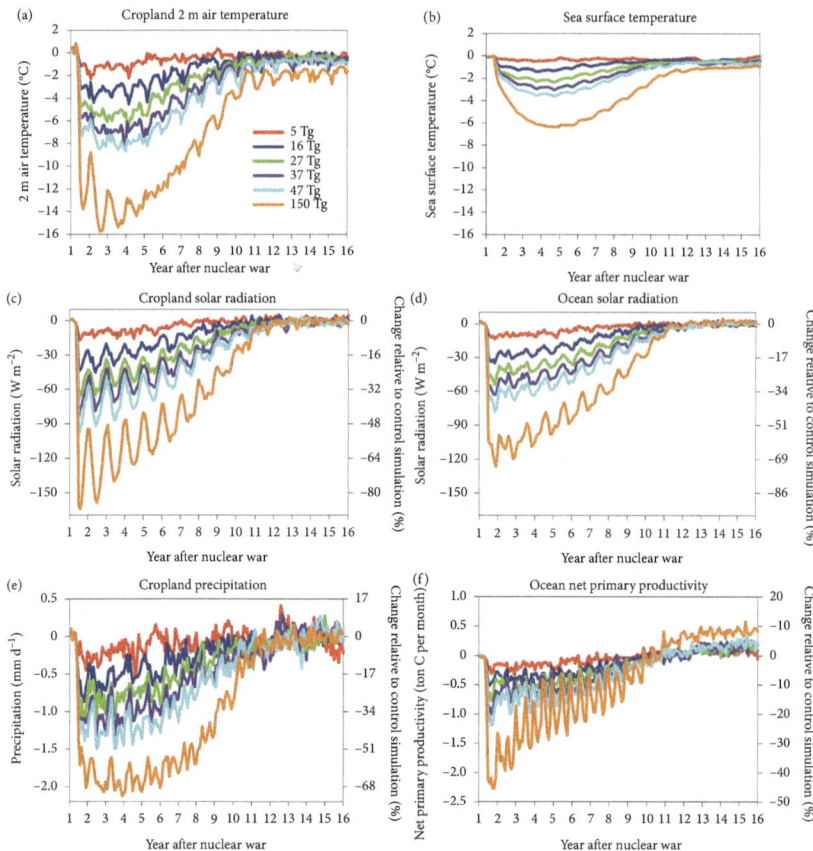

Figure 13.2 Climatic impacts by year after different nuclear war soot injections. a–f, changes in cropland surface temperature (a), solar radiation (c), and precipitation (e) averaged over global crop regions of 2000, and sea surface temperature (b), solar radiation (d) and net primary productivity (f) averaged over the oceans following the six stratospheric soot-loading scenarios in Table 11.1 for 15 years following a nuclear war. These variables are the direct climate forcing for the crop and fishery models. The left y-axes are the anomalies of monthly climate variables from simulated nuclear war minus the climatology of the control simulation, which is the average of 45 years of simulation. The right y-axes are the percentage change relative to the control simulation. The wars take place on May 15 of Year 1, and the year labels are on January 1 of each year. For comparison, during the last Ice Age 20,000 years ago, global average surface temperatures were about 5°C cooler than present. Ocean temperatures decline less than for land because of the ocean's large heat capacity. Ocean solar radiation loss is less than for land because most ocean is in the Southern Hemisphere, where slightly less smoke is present.

From Xia et al. (2022)

around 3 mm/day as smoke increases because there is less energy available to evaporate water from the oceans and the land. Sea surface temperatures (Figure 13.2b) drop by less than cropland temperatures (Figure 13.2a) because the ocean has more mass than the land due to sunlight penetration 100 m into the oceans. Solar energy only penetrates centimeters into the soil daily, and meters into the soil on an annual basis. Still, the declines in ocean surface temperature, along with losses of sunlight over the ocean (Figure 13.2d) are large enough to impact the net primary productivity in the oceans (Figure 13.2f). The primary productivity is the rate at which the life in the seas, mainly phytoplankton, produces organic carbon, which is food for the animals in the sea. While primary productivity is reduced 40% for the 150 Tg war, it went to almost 0 for the K-Pg extinction event because there was not enough light for photosynthesis to occur for several years, as shown in Figures 4.1 and 12.6.

Using the CLM5crop model, Xia et al. (2022) computed the loss of calories due to the changes in the climate parameters, as shown in Figure 13.3. The global average calorie losses due to losses in crops are shown in Figure 13.3a. Relative to the results of Jägermeyr et al. for the 5 Tg case, the CLM5crop results are slightly smaller. These losses increase dramatically with the increases in soot, until almost 90% of calories are lost for a war between Russia the United States and their allies. Several factors are not included in these calculations that might make the results worse, such as increased surface UV light due to loss of the ozone layer and loss of fuels and vehicles needed to farm.

If it will be very difficult to grow food on land, could fish supply a portion of global food needs? Fish provide only a small fraction of calories today, and Figure 13.3b suggests that marine fisheries would also be depleted. Therefore, agricultural losses caused by nuclear war are unlikely to be offset by the world's fisheries, particularly given widespread overfishing today (Scherrer et al. 2020). Cold temperatures and reduced sunlight would decrease the productivity of fish, and although intensified fishing efforts could yield a small short-term increase of fish catch, those catches would fall dramatically within a few years due to overharvesting.

We also discovered that a "Nuclear Niño" would follow nuclear war (Coupe et al. 2021). El Niño and La Niña are irregular variations in the sea surface temperature of the Pacific Ocean that have major impacts on tropical fisheries. Following nuclear war, there would be a super El Niño that would last for 5 years or more. In El Niño conditions, upwelling of water along the equatorial western coastlines of the Pacific Ocean is diminished. Upwelling water is rich in nutrients for fish, and its loss will damage major fisheries. We found that reduced sunlight and changes to ocean circulation would result in up to a 40% reduction in equatorial Pacific phytoplankton productivity, and these phytoplankton are at the base of the oceanic food chain. That, combined with toxic and radioactive

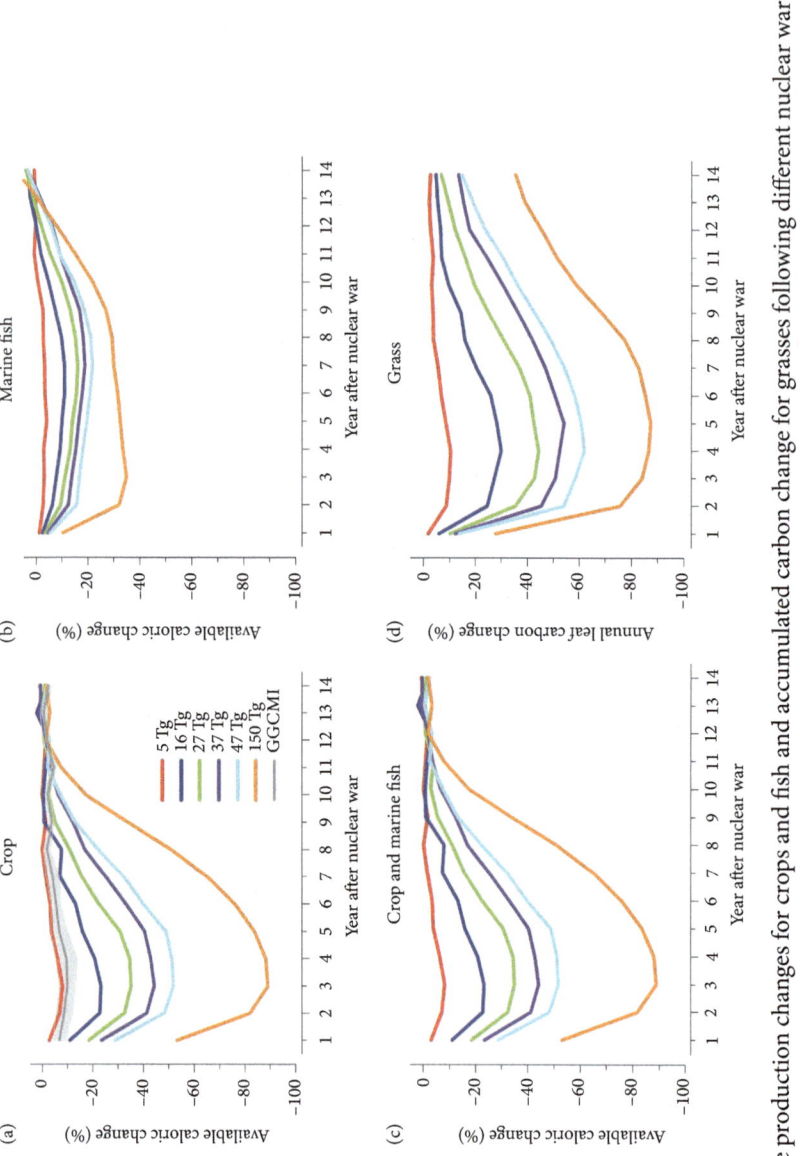

Figure 13.3 Calorie production changes for crops and fish and accumulated carbon change for grasses following different nuclear war soot injections. a) global average annual crop calorie production changes (%; maize, wheat, rice and soybeans, weighted by their 2010 observed production and calorie content), b) marine fish production changes, c) combined crop and fish calorie production changes, and d) grass leaf carbon, a combination of C3 and C4 grasses, and the change is calculated as annual accumulated carbon. For context, the gray line (and shaded area) in (a) are the average (and standard deviation) of six crop models from the Global Gridded Crop Model Intercomparison (GGCMI) under the 5 Tg scenario. CLM5crop shows a conservative response to nuclear war compared with the multi-model GGCMI response (Jägermeyr et al. 2020).

pollution, would severely limit the food sources in the oceans. Furthermore, the large temperature contrasts between the oceans and the land would produce strong storms that would make fishing difficult at best. This means that any survivors along coastlines would have a difficult time feeding themselves by fishing or harvesting seaweed.

13.4 How Many Could Survive Starvation after a Nuclear War?

Xia et al. (2022) translated the loss of calories into the number of people who might survive through the second year after the nuclear war. At that time, reserves of food should have been exhausted. Unfortunately, calorie losses extend well beyond the second year; so, even more people are likely to perish in following years. There are many assumptions needed to predict survival. For instance, there are already parts of the world, mainly in Africa and the Middle East, but also in parts of South America and Asia, where individual countries do not produce enough food currently to feed their populations the basic numbers of calories needed to function. These countries depend on international transportation of food for their populations to survive. However, it is likely that after a nuclear war that transportation would largely stop, either because the infrastructure (refineries, ports, vehicle manufacturing plants) was destroyed or because countries are hoarding food for their own citizens.

The average person on Earth now consumes about 2,010 calories per day. People will lose weight if they eat only 1,911 calories on average, and they need 1,145 calories per day to maintain just their resting energy expenditure. If faced with crop losses and assuming people keep their livestock and fish, there would not be enough food for everyone to maintain their weight after a war with 16 Tg of soot. And after the biggest India-Pakistan war with 47 Tg, people would not have enough food on average to maintain life unless food waste were greatly reduced. Of course, following a nuclear war people might choose to eat the grain they were feeding their livestock. The greater fraction of livestock that is retained, the fewer the calories available to people in most countries. Even then, a war between Russia, the United States, and their allies would mean that the global population could not maintain itself.

The results paint a grim picture, with billions of people dying even from 16 Tg of soot injected. Nuclear war would impact everyone on the planet, not just the millions or hundreds of millions of people killed by the nuclear explosions in their countries. No matter how much waste is eliminated, and no matter how much livestock is eliminated, a war between India and Pakistan, which might kill as many as 125 million locally, is likely to place 1 to 3 billion people worldwide in danger of starvation. The range of deaths depends on the yield of the weapons used, which determines the amount of smoke produced. In the case of a war between Russia, the United States, and their allies, more than 5 billion people would die out of the 6.7 billion assumed to be living at the time of this calculation.

Table 13.1 contrasts the deaths in a countervalue war due to blast, radiation, and fires with the deaths due to agricultural failure. More than 10 times as many people could die from starvation than would die from the horrific, direct effects of nuclear war. More than 90% of the deaths are likely to be due to fires, climate change, and agricultural failure—effects that the military considers to be collateral or accidental deaths accompanying the fatalities they do account for when destroying their true targets. We do not have the information to construct a similar table for a counterforce war because at the time of this writing there are no evaluations of the smoke or direct fatalities in a counterforce war.

Figure 13.4 illustrates the fraction of the populations surviving starvation by the second year following the war. We assumed perfect distribution systems within countries, that each person who will survive eats the minimum needed for survival, and that the others got no food. Of course, such a situation would lead to open revolt, likely leading to even a higher fraction of people dying.

Table 13.1 Number of Weapons on Urban Targets, Yields, Direct Fatalities from the Bomb Blasts, and Resulting Number of People in Danger of Death Due to Famine for the Different Scenarios Studied by Xia et al. (2022)

Soot (Tg)	Number of Weapons	Yield (kt)	Number of Direct Fatalities	Number of People without Food at the End of Year 2[a]
5	100	15	27,000,000	255,000,000
16	250	15	52,000,000	926,000,000
27	250	50	97,000,000	1,426,000,000
37	250	100	127,000,000	2,081,000,000
47	500	100	164,000,000	2,512,000,000
150	4,400	100	360,000,000	5,341,000,000
150[b]	4,400	100	360,000,000	5,081,000,000[b]

[a]The last column is the number of people who would starve by the end of Year 2 when the rest of the population is provided with the minimum amount of food needed to survive, assumed to be a calorie intake of 1,911 kcal per capita per day, and allowing for no international trade; 50% of livestock grain feed is used for human consumption, and 50% of livestock grain feed is used to raise livestock. For 2010, the total population of the nations used in this study was 6,700,000,000. Currently, it stands at 8 billion, so even more deaths would occur now. There are many other scenarios in which these amounts of soot could be produced by a nuclear war, and the scenarios we use are only meant to be illustrative examples. The last column is the case with the fewest number of deaths without international trade, and other cases are available in Xia et al. (2022).

[b]Assuming total household waste is added to food consumption.

Proportion of population that would starve to death
Partial Livestock Case, 37 Tg, livestock feed to human consumption, no trade

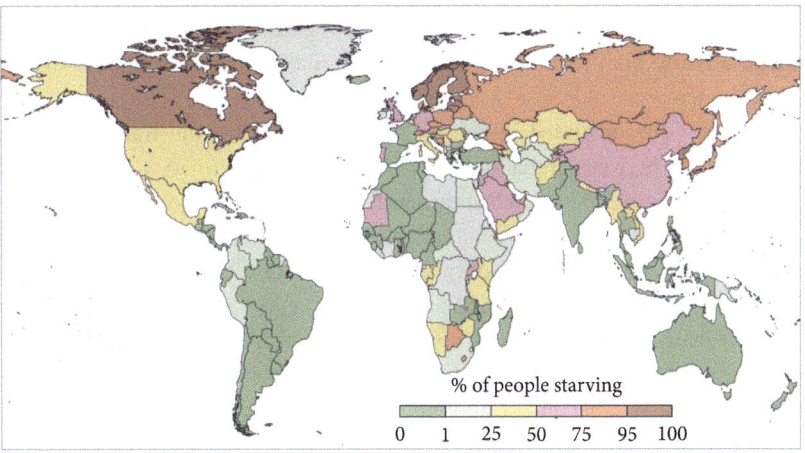

Proportion of population that would starve to death
Partial Livestock Case, 150 Tg, 50% livestock feed to human consumption, no trade

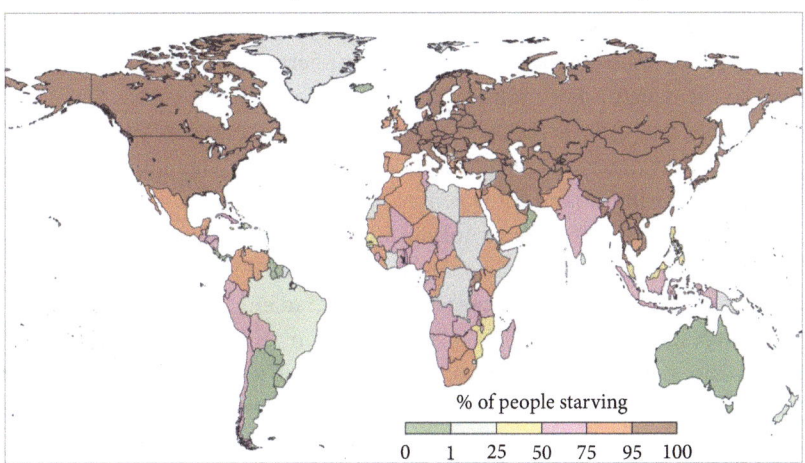

Figure 13.4 National population (%) surviving at the end of Year 2 for a calorie intake of 2,200 kcal/capita/day when the rest of the population is provided with no food, and for no international trade. Top figure is for a war between India and Pakistan in which 50% of the livestock food is used for humans, and 37 Tg of soot is generated by the explosions of 250 nuclear weapons with yields of 100 kt. Bottom figure is for a war between the United States and Russia and their allies, with 150 Tg of soot produced and 50% of livestock food used by humans.
From Xia et al. (2022)

The figure illustrates deaths through the second year after the war, but conditions would not improve for several more years, leading to further deaths. These assumptions maximize the survival numbers. Whether people chose to become vegetarians and not maintain livestock did not make much difference to overall survival rates.

Two cases are illustrated in Figure 13.4. The top panel illustrates the 37 Tg case for a war between India and Pakistan assuming that 250 weapons with a yield of 100 kt are used on urban targets (Table 11.1). The bottom panel illustrates the results of a war between the United States, Russia, and their allies with 150 Tg of soot. In both cases, only a few percent of the population of Russia survives the second year, and similar losses occur in Canada if they maintain half their livestock. High-latitude countries are highly vulnerable because they already experience short growing seasons due to their latitude. If temperatures fall below normal, it becomes impossible to grow crops. Few countries in northern midlatitudes have more than 25% of their population surviving, also because of the shortened growing season in a nuclear winter. Countries in the Southern Hemisphere do not suffer as much, with 50% or more of their populations surviving in most countries. There is less death in the Southern Hemisphere because there is less smoke in the Southern Hemisphere, the large amount of ocean moderates the temperature, and the Southern Hemisphere countries are closer to the equator. For example, the border between the United States and Canada generally follows the 49°N parallel, while the Straits of Magellan, near the tip of South America, are at a latitude of about 53°S. Therefore, there is no Southern Hemisphere land equivalent to Canada or to Russia.

Many additional factors have not been included in the analysis of Xia et al. (2022). Large global decreases in ozone have not been observed in human experience, though there is a significant latitudinal gradient in ultraviolet intensity. It is known, for instance, that the incidence of skin cancer in people with light-colored skin increases with decreasing latitude because of increasing ultraviolet light (Fears et al. 1976). The ultraviolet light is greatest in the tropics because the sunlight is more intense there and because there is less ozone in the tropics than elsewhere. While we know that enhanced ultraviolet light is hazardous, its effects are not yet included in agricultural models, so it is not yet possible to gauge its impact on food supplies or the environment. Current calculations show that the ozone loss is mostly caused by soot heating the stratosphere (Mills et al. 2008, 2014; Bardeen et al. 2021). Earlier studies focused on nuclear fireballs from large bombs penetrating up into the stratosphere. The intense heat of a bomb blast burns the atmosphere, which is mainly composed of O_2 and N_2, producing NO and NO_2 (NO_x), which act as a catalyst to destroy ozone. However, there are now fewer large weapons, and their impact on nitrogen oxides is therefore reduced. Calculations show reductions of ozone in the stratosphere by as much as 50%, with the effects only gradually ameliorating over a period of years. A trend toward smaller nuclear weapons lessens the number of fireballs reaching the stratosphere, but the changed atmospheric circulation still takes some NO_x up into the ozone layer and into the Southern Hemisphere, and the stratosphere would be so hot due to heating by smoke that ozone-destroying reactions would proceed much more rapidly, producing a global "nuclear winter ozone hole" that would last for years. Initially, for large smoke amounts the smoke would absorb the enhanced UV, but in several years, more would get to the surface, affecting humans and other organisms. There would be increases in sunburn, skin cancer, and cataracts, as well as large impacts on crops and phytoplankton.

Another factor not included in Xia et al.'s calculations is that temperature declines caused by nuclear war would lead to carbon dioxide dissolving into the oceans, which in turn would affect ocean acidification and adversely impact the ability of marine calcifying organisms to maintain their shells and skeletons in a corrosive environment (Lovenduski et al. 2020). We have no experience with the impacts of these changes on the global biota. In addition, cold temperatures would lead to expansion of sea ice, which could make it more difficult to fish from high-latitude ports in Japan, Iceland, and Alaska, as shown by Harrison et al. (2022). After a nuclear war, temperature changes and pollution would affect the fish in the ocean. Because the land would cool more than the ocean, the temperature contrast would generate stronger coastal storms, which would affect fishing boats. Harrison et al. also show that the nuclear cooling would produce expansion of sea ice into populated coastal areas and decimation of ocean marine life. Marine ecosystems would be highly disrupted, resulting in impacts to ecosystem services worldwide and lasting for decades.

We have a lot of experience with the effects of changes in temperature and precipitation on agriculture, and those feed into the crop modeling results described above. The crop declines are mainly due to reduced temperatures and shortened growing seasons, but these crop declines do not consider social feedbacks. For instance, farmers may abandon their land if they think it is poisoned by radiation or may concentrate on feeding their local communities rather than exporting food. Harwell and Hutchinson (1985) provide a detailed discussion of the effects on farming of transportation breakdowns, difficulties in obtaining fuel for farm machinery, disruptions to grain markets, and economic challenges such as obtaining loans for planting food (and many other issues). These social responses can be extremely important to the food supply, possibly even more important that the direct effects of climate on agriculture.

Another factor in food supply is the disruption of the distribution mechanisms. Not only would it be virtually impossible to grow food for several years after the nuclear holocaust it would also be impossible to obtain food from other countries. It is hard to imagine that countries with some food production (possibly Argentina, Australia, or New Zealand) would be willing to export it in such an uncertain international situation. Trade in fertilizer, herbicides, pesticides, and genetically altered seed would also be disrupted. For a regional nuclear war with less cooling, such as between India and Pakistan, but with still important disruptions of food production, these market disruptions could produce more shortages than the direct impacts on agriculture. In addition to the disruption of food, there would be many other stresses for any surviving people. These would include the lack of medical supplies and personnel, high levels of pollution and radioactivity, psychological stress, rampant diseases and epidemics, and increased ultraviolet radiation when smoke does not significantly block light transmission. In addition, it is likely that residual warfare involving militias and surviving military people would occur.

The tremendous productivity of the grain belt of the United States and Canada feeds not only those countries but also many in the rest of the world, occasionally including Russia, where normal climate variability often results in reduced harvests. This productivity is the result of modern farming techniques that allow less than 2% of the population to produce more than enough for the rest. To do this, tremendous energy subsidies are needed. Farmers depend on fuel for their machinery, fertilizer, and pesticides, none of which would be available or distributed in the aftermath of a war. Furthermore, insects have a higher tolerance for radiation and the stresses that would follow a war than do their predators such as birds. Whatever might grow would be eaten by pests, already a significant problem in today's production. Also, the seeds in use were designed to yield high productivity assuming the current climate and inputs of chemicals and energy as discussed above. These seeds would not thrive in a radically altered growing environment. Our dependence on technology is such that if every human in the United States went out to the fields to try to raise crops with manual labor, and if they knew what they were doing, and if they had enough food to eat, and if they were healthy, they still could not produce what is grown today. Estimates from the 1980s according to Harwell and Hutchinson (1985) suggest that humans using primitive agricultural techniques and lacking modern social systems could only support a global population of a few hundred million.

The effects of a nuclear conflict on health would compound the misery. Immune deficiencies can be produced by burns and trauma, radioactivity, malnutrition, psychological stress, and ultraviolet radiation once soot no longer absorbs the ultraviolet light. All of these would be present for the survivors in the target nations. Pollution from dioxins, PCBs, asbestos, and other chemicals from urban fires will make the air unhealthy to breathe. Severe psychological stress will prevent some of the survivors from making the efforts to continue to exist. In addition, in countries in which large segments of the population were killed, decaying bodies would likely spread disease. Moreover, much of the world's medical facilities would be destroyed, as analyzed for the United States by Helfand et al. (2002).

PART III
EPILOGUE. COULD IT HAPPEN?

Chapter 14

Will Humans Become Extinct from an Asteroid Collision or a Nuclear War?

About 75% of known species became extinct when an asteroid hit Earth 66 million years ago. As we discussed in this book, the same phenomena that caused the K-Pg extinction are likely to occur following a nuclear war. Extinction of our species was not ruled out in initial studies of nuclear winter by biologists (Ehrlich et al. 1983). It is useful to consider how we might predict the extinction of a species, as well as how asteroid collisions, nuclear wars, and other phenomena might cause ours.

14.1 The Nature of Extinctions

Species come and go. None lasts forever. The geological record is replete with the appearance of new life forms and the disappearance of older ones. Most species vanish for no apparent reason as random variations of DNA produce slight differences in siblings. Some of these differences allow a competitive advantage to the younger creatures over older life forms, which slowly fade away. Other species come and go in response to environmental changes. The building of mountains, the formation of seas, the evaporation of oceans, and the movement of continents to new locations may all trigger the death of creatures that could not tolerate their new environment, and it might allow the rise of fresh species that could take advantage of the opportunities.

Occasionally, global catastrophes produce a worldwide shock to the environment that can exterminate many life forms across the planet, allowing surviving creatures that were not competitive in the old environment to inherit the Earth. Such an event led to the extinction of the dinosaurs and allowed our mammalian ancestors to finally take over the planet, which they had inhabited for nearly as long as the dinosaurs. It may be only a smug conceit to think that something similar will not happen to us in the not-too-distant future. A large-enough asteroid or comet impact in the future, while unlikely, would kill most life on Earth. A large-enough comet to produce mass extinctions if it hit Earth is approaching the inner solar system as we write this book, but it poses no threat to us since it

Earth in Flames. Owen Brian Toon and Alan Robock, Oxford University Press. © Oxford University Press (2025). DOI: 10.1093/9780197799734.003.0014

will not reach Earth's orbit. What about a nuclear war? We need to understand mass extinctions and other factors that might cause humans to become extinct.

14.2 Mass Extinctions

Extinctions are a constant occurrence in geologic history as species originate and then either disappear or are replaced by evolutionary offshoots. Mass extinction events are geographically widespread losses of diversity and abundance of organisms that are rapid compared to the duration of the organisms in the geologic layers. Unfortunately, we have no record of most organisms. Instead, we know about those organisms that are most likely to be preserved as fossils and are numerous enough and large enough for geologists to have a good record of their history. Most of the record we have of extinction events is based on marine fossils, because when marine organisms die they settle to the bottom of the sea and build up layers of sediment. Unfortunately, the ocean floors are relatively young, much of it formed after the extinction of the dinosaurs, because continental drift is constantly erasing old ocean floor by subduction at the continental margins, adding new crust at the mid-ocean ridges. The US National Geophysical Data Center maps show that the oldest ocean floor, about 200 million years old, is in portions of the western Pacific and the northwest Atlantic. Parts of the Mediterranean Sea may be even older. While these are long-time scales compared to a human lifetime, they are less than 5% of the history of life on Earth. Fortunately, some old sea floor has been preserved as part of what is now a continent. The land is a poor source of fossils because of erosion, so most fossils are lost as land is eroded away. Usually, those land-based fossils that are found have been swept into a swamp, river, or lake and jumbled up with others. Occasionally, volcanic dust will suddenly bury creatures on the land and preserve them, or an ancient swamp may be preserved as a coal bed.

The biological world is classically divided into a series of ever more closely related groupings (Figure 4.1). It is more difficult to cause the extinction of a higher-level group than a lower-level one because the higher-level group has greater diversity. For instance, mammals form a class in the biological organization scheme. A class contains many orders, which in turn contain many families. The families contain numerous genera that may be made up of many species. Humans are a species, *Homo sapiens*, the only member of the genus *Homo* still in existence, in the family Hominidae, in the order of Primates. Our family includes our genus, and those of chimpanzees (*Pan*), gorillas (*Gorilla*), and orangutans (*Pongo*). It is thought that our genus split from that of the chimpanzees between 4 and 8 million years ago. There have been many species that were members of our genus in the past. In the recent history of our genus, covering perhaps

200,000 years, Neanderthals, *Homo floresiensis*, and *Homo denisova* coexisted with us *Homos sapiens*. There were also many earlier species classified under the genus *Homo*.

John Sepkoski Jr. (1996) studied extinction in a database for marine animal genera that included vertebrates (creatures with backbones such as fish); invertebrates (creatures without a backbone such as sponges, mollusks, crabs, jellyfish, and worms), which are by far the dominant number of species of animals; foraminifera (generally microscopic, single-celled animals whose shells are largely made of calcium carbonate), which are so abundant that they form much of the seabed in the deep oceans; and radiolaria (generally microscopic, single celled animals whose shells are siliceous), which also compose much of the seabed. The database contained 33,180 animal genera. The fossils of multicellular organisms first become easy to find from about 600 million years ago, when shells or other body parts that did not decompose rapidly first developed, probably in response to the onset of predation. The data on genera (Figure 14.1) show a slow increase over time in the number of genera that can be recognized in the fossil record in the past 600 million years, with about four times as many genera now as 400 million years ago. Possibly the increase in genera is due to actual diversification of the biota, or possibly it is just that more sediments are exposed for more recent times, allowing more genera to be identified.

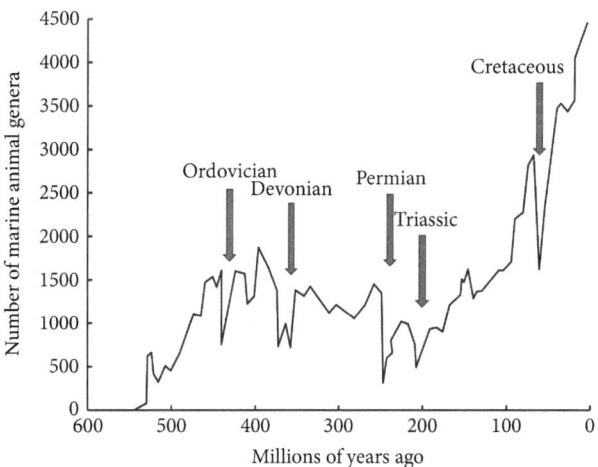

Figure 14.1 The history of the number of marine animal genera based on the work of John Sepkoski Jr. Arrows mark the five great extinctions. The first mass extinction occurred at the end of the Ordovician, the second in the Late Devonian, the third at the end of the Permian, the fourth at the end of the Triassic, and the fifth, the dinosaur extinction event, at the end of the Cretaceous.

Figure 14.1 also shows that there were five great mass extinction events in the fossil record, when the number of genera suddenly declined. There are also many lesser extinction events affecting smaller number of genera. At the K-Pg boundary, when the dinosaurs vanished, 47% of the genera studied became extinct, and 16% of families vanished.

Extinctions of species are more difficult to track in the fossil record than those of families or genera, in part because many species do not last very long. David Jablonski (1994) found that 90% of species went extinct outside of times with mass extinction events. However, extrapolating from Sepkowski's estimates of losses of genera, he estimates that 75% of the living species in these genera went extinct at the K-Pg boundary. These two percentages might seem contradictory at first, but they simply imply that only a small fraction of all the species that have ever lived were present at the time of the K-Pg extinction. In general, it is thought that at any given time only a few percent of all the species that have ever lived are still living. It is not possible to give a meaningful lifetime for the average species because there is a wide range of variation. For example, horse-shoe crabs and their ancient relatives, common inhabitants of sandy seashores today, have been around for 450 million years. Nevertheless, estimates have been made of the average time before a species goes extinct for various organisms, such as marine invertebrates (5–10 million years), mammals (1 million years), and marine animals (4 or 5 million years) (Lawton and May 1995). The short durations of species lifetimes may suggest that we *Homo sapiens* will likely not be present on Earth for long. Assuming we have the lifetime of a typical mammal species, the 200,000-year record of *Homo sapiens* might be 20% of our duration as a species. Because we are the first species to be aware of the short duration of species, we may also have the power to control our own fate and extend our species lifetime. However, there are many possible fates for our species, some of which suggest we will soon be replaced (or at least no longer be the only species in our genus).

14.3. The Likely Fate of *Homo sapiens*

A probable fate is that our species will be replaced by many newer, related species. During the coming decades and centuries, it is likely that humans will colonize the local solar system, especially the moon and Mars, which have much lower gravity than Earth. Flights of humans on the Space Station and other long-duration vehicles show that the human body responds rather quickly to lowered gravity. Bone density declines at about 1–2% per month. Blood mass, muscle mass, and strength are all measurably reduced in flights lasting only a few weeks. Fluids are also redistributed, leading to "puffy face-bird leg" syndrome (Williams et al. 2009). It is possible that astronauts on very long duration flights, such as to Mars, could have bone loss that cannot be replaced entirely on return to Earth.

While people may be able to freely go to planets with lower gravity, it is possible that they cannot easily go to ones with higher gravity. People born on Mars or the moon might not be able to travel to Earth due to the physiological changes, such as reduction in bone strength, caused by lower gravity where they were born. In the distant future relative to a human lifetime, our ancestors may move to planets around other stars. The travel time to these planets might be many human lifetimes, and quick return would be unlikely. Eventually, such isolation will lead to the evolution of new species in the genus *Homo*, just as isolation has previously led to new species of other animals and plants. The famous bar scenes in *Star Wars* films, in which humans and aliens get together to share food, drink, and breathing the same atmosphere, may actually be gatherings of various future species in the genus *Homo* that developed on different planets seeded by wandering Earthlings.

Science fiction writers have long envisioned another possible fate for *Homo sapiens*: to be replaced by robots. Of course, this is not evolution in the normal sense. Some scientists, as well as leading developers of technology and computation, have warned that this fate could be realized in the near future. There is plenty of evidence that the first wave of this replacement is well underway and causing significant problems for humans. Computer-driven machines are displacing people in many industries, permanently eliminating jobs in manufacturing and other areas. Most of these machines do not have to "think" or make decisions; they simply perform highly precise, repetitive tasks. However, as computers become more capable, and we learn more about artificial intelligence (AI), computers will make more decisions. For instance, we are at the beginning of the era of self-driving cars. These cars will eventually displace vast numbers of taxi drivers and truck drivers from their jobs. They may also alter car ownership by allowing us to have a car at our beck and call, instead of needing to have one sitting idle in our garage waiting for us to use it. As these jobs for humans are eliminated, it will free people for much more creative pursuits, but it may also make our current population levels unnecessary to support us. Recent developments in AI already make it difficult to distinguish speaking and writing done by computers from that done by people. As university faculty, we had been concerned about whether a student wrote the term paper the student turned in or whether it was taken off the web. Now we will need to worry whether an AI program might have written the term paper. Brian tells his students to use AI if they wish, but to let him know if it was useful.

14.4 The Asteroid Impact and the Development of New Species

Not all the dinosaurs went extinct in the K-Pg cataclysmic event. In fact, the descendants of flying dinosaurs are all around us as chickens, turkeys, hawks, eagles, and other birds. However, the dinosaurs were never again the dominant life forms on Earth. Dinosaurs originated in the Triassic Period about 230–245 million years ago after the largest known mass extinction event ended the Permian Period, but their reign as rulers of the Earth started after yet another mass extinction event that began the Jurassic geologic period about 200 million years ago. So, dinosaurs themselves benefited from mass extinctions opening opportunities that they could exploit. Mammals first appeared at about the same time as dinosaurs. While dinosaurs dominated the larger animals on the planet, mammals, reptiles and amphibians dominated the smaller-sized animals. The reign of mammals as rulers of the larger animals on Earth did not begin until the start of the Paleogene geologic period, when the asteroid impact eliminated the non-avian dinosaurs. It was the impact that eventually led to humans by opening previously occupied niches that mammals could exploit.

Mammals did not outcompete the dinosaurs to take over Earth; we simply survived a catastrophe that the dinosaurs could not. Without the asteroid impact, dinosaurs might still rule the Earth. Of course, they would have evolved considerably in the past 66 million years.

Paleontologist Dale Russell forecast forward from the final days of the dinosaurs to predict what the dinosauroid rulers of Earth might now look like if the impact had not occurred. He started with the *Stenonychosaurus inequalis*, a dinosaur with one of the largest brains relative to its body size, which is often used as a criterion for intelligence. The ratio of its brain mass to its body mass was about the same as a modern ostrich's. In other words, the *Stenonychosaurus inequalis* was probably not too bright.

Stenonychosaurus inequalis were turkey-sized, birdlike dinosaurs. Figure 14.2 illustrates a model of a dinosauroid built by Dale Russell and Co-imaginer Ron Séguin (1982). They thought that if *Stenonychosaurus inequalis* evolved to have a brain size like modern humans then the need to support the mass of the head would require it to lose its tail and walk upright. It might retain the three fingers of *Stenonychosaurus inequalis* and their binocular vision and large eyes. In the early 1980s, before we knew that many dinosaurs were feathered, Russell and Séguin imagined the creature might be green with a reptilian skin. Unlike *Stenonychosaurus inequalis*, they imagined the creature would give live birth to accommodate the large head, so it would have a navel. Russell and Séguin developed the features of the dinosauroid based on deductive reasoning. However, the dinosauroid looks a lot like a human self-portrait. Others have since argued that the form of evolved dinosaurs would depend on their evolutionary history,

Figure 14.2 Model of a dinosauroid built by Dale Russell and Co-imaginer Ron Séguin, Russell and Séguin (1982). The replica creature in the background is based on actual fossils of *Stenonychosaurus inequalis* found by Russell. The creature was about as large as a turkey. The dinosauroid might have eventually ruled the Earth had the impact not occurred 66 million years ago. Would dinosauroids have created nuclear weapons?

Copyright Canadian Museum of Nature

and there is nothing that requires large-brained intelligent creatures to look like people rather than large feathered birds with long tails. Either way, it is our good fortune to be here rather than the dinosauroids.

While the non-avian dinosaurs are the most famous group of creatures to go extinct at the K-Pg boundary, many other life forms also vanished. For instance, while birds survived, the pterosaurs, large flying reptiles, did not. Many species of mammals, birds, insects, lizards, and plants also went extinct. In fact, only about 7% of mammal species and 16% of mammal genera survived the extinction event (Longrich et al. 2016). The extinction was just as severe in the ocean as the land. Large marine reptiles such as mosasaurs and plesiosaurs vanished. While shelled creatures such as the nautilus survived, the similar ammonites that are more closely related to octopus, cuttlefish, and squid did not. Many fish, sharks, and plankton also vanished.

These losses of species, many with long histories, were followed by rapid evolution of new species. What might happen following a nuclear war?

14.5 Do Astronomers Have Evidence That Intelligent Life May Be Destroyed in Nuclear Wars?

Fermi's paradox is a famous problem in astronomy. Why is there no conclusive evidence of extraterrestrials despite the likelihood that they exist? Enrico Fermi, a Nobel Prize–winning physicist, uttered a phrase at a lunch at the Los Alamos National Lab in 1950 after a discussion about UFOs that was something like "where is everybody?" Fermi was hardly the first to speculate on the existence of life elsewhere in the universe, but he never published anything about why extraterrestrials have not been discovered.

As one answer to the paradox, astronomers Iosif S. Shklovskii and Carl Sagan (1966) drew attention to the possibility that intelligent life may have a short lifetime. Carl later expanded on this idea in his book and television series, *Cosmos* (Sagan 1980). They argued that vast numbers of intelligent civilizations should have developed in our Milky Way Galaxy and that their failure to contact us suggests they have not survived. They attributed this short lifetime to civilizations inadvertently committing suicide via nuclear wars soon after they arose.

At the height of the Cold War, Shklovskii—a noted Soviet scientist, and Sagan, an up-and-coming American scientist, collaborated on a book titled *Intelligent Life in the Universe*. This book is remarkable because it broke down an incredibly complex, seemingly unapproachable subject into little pieces that could be tackled. The authors of the book also provided a significant demonstration that science transcends national boundaries, even when those boundaries divide fundamentally different views on many subjects.

Shklovskii and Sagan used an equation famously attributed to Frank Drake to estimate the number, N, of advanced technical civilizations in our galaxy that might be inclined and able to communicate across the Milky Way at the present time. Many papers have been written about variants of the equation and what the proper values of the terms might be. Despite its defects, the equation does provide a systematic way to think about the origin and ends of intelligent life on Earth and in the universe. This equation reads as follows:

$$N = N^* f_p n_e f_l f_i f_c f_L$$

where N^* is the number of stars in our galaxy; f_p is the fraction of stars with a planet; n_e is the number of planets in a given planetary system that are ecologically suitable for life; f_l is the fraction of planets on which life arises; f_i is the fraction of planets with life on which intelligent life has arisen; f_c is the fraction of planets with intelligent life that has released signals that could be detected across the galaxy; and f_L is the fraction of the planet's lifetime during which the civilization has communicated.

N^*, f_p, and n_e have been crudely measured by astronomers in the last few years. Multiplying these numbers together yields the number of planets in our Milky Way Galaxy that should be suitable for life. If we assume that f_l is 1, based on the short time it took for simple life to arise on Earth, then there are about 2×10^{11} stars in the Milky Way with planets that have had life originate on them. Nearly all the nearby stars with planets should have been home to life at some point in their histories. It is hard to imagine what a number like 2×10^{11} means. The population of the Earth is about 8×10^9 people, so each of us has about twenty-five stars with simple life forms in the galaxy for our own if we could only get there. The Obama administration, in exchange with Congress for a nuclear treaty with Russia, agreed to allow the nuclear arsenal of the United States to be refurbished. The estimated cost of this is at least US$1 trillion (or $5 for every inhabited star in our Milky Way Galaxy).

Shklovskii and Sagan argued that intelligent life is inevitable given enough time because it confers important survival strengths; they also argued that intelligent life will learn to communicate (and so both f_i and f_c are 1), implying that every planet with life that stays long enough in the Habitable Zone will develop intelligent life. There should be around 2×10^{11} planets, almost a trillion planets, in the Milky Way that developed intelligent life. Where is everybody? One can also use this argument backward. If there are 2×10^{11} planets on which life arose, then there would be at least one planet in the Milky Way, Earth, with intelligent communicating life if the odds of intelligent life developing and learning to communicate is greater than five in a trillion.

Now we finally come to the last term in Drake's equation, f_L, the fraction of the planet's lifetime during which the civilization has communicated. If f_L is close to 1, the Milky Way should be teeming with civilizations eager to communicate with us. If, however, our 100 years with radio waves leaking into space that could be heard by life on other stars is typical of the duration of intelligent life, and societies destroy themselves after that time, then f_L may be about 2.5×10^{-8}, and there may be only 5000 advanced communicating civilizations in our galaxy instead of 2×10^{11}. Thus, Shklovskii and Sagan's solution to the Fermi paradox is that we would have to listen for radio broadcasts from about 40 million stars before we heard one sending us signals.

There have been many alternative suggestions to resolve Fermi's paradox rather than a short lifetime of intelligent civilizations. Peter Ward and Donald Brownlee (2000) outline some of these ideas. Perhaps it is very difficult for intelligence to arise, but the universe is teeming with single-celled organisms. Perhaps intelligent civilizations don't want to be detected and so they don't communicate. Perhaps large asteroid impacts, gigantic volcanic eruptions, supernova explosions, and other deadly phenomena extinguish life in other solar systems. Unfortunately, we don't know enough at present to evaluate the likelihood of these other explanations of the Fermi paradox.

The Drake equation cannot yet tell us the number of intelligent civilizations because the factors that make up the equation are too poorly known. It is likely that satellites currently on the drawing boards or under construction will detect life on other planets in the not-too-distant future, and we will then be able to determine more of the uncertain factors in the equation. Many factors might explain a low number of intelligent civilizations other than nuclear annihilation. Nevertheless, we should take Shklovskii and Sagan's warning seriously that the lifetime of civilizations may be short and do what we can to avoid this fate for our world.

14.6 Survival in the Aftermath of a Nuclear War

The non-avian dinosaurs became extinct 66 million years ago due to the impact of an asteroid near modern-day Chicxulub, Mexico. The phenomena that followed—massive fires, darkness at noon, subfreezing temperatures, and severe damage to the ozone layer—are all phenomena that we expect after a nuclear war. Should we expect the extinction of humans after a nuclear war?

We first need to note that the Chicxulub impact was a very extreme event. The energy of the impact, which we know from the size of the crater and the amount of iridium in the K-Pg layer, was about 100,000 times greater than would be released from use of the current nuclear arsenals. The amount of black carbon that entered the atmosphere 66 million years ago, and still remains in the geologic layer that marks the event, is about 100 times greater than is likely to occur from a nuclear war. From calculations, we expect that light levels after the impact were less than 1% of normal, below the limit of photosynthesis, for several years. We do not expect light levels to fall below 20% following a nuclear war. Therefore, collapse of the oceanic food chain due to cessation of photosynthesis is not as obvious following a nuclear war as for the K-Pg event. Temperatures following the impact drop to similar levels as expected for a nuclear conflict, which would be very damaging to agriculture. Looking at the past, we know a mass extinction occurred following the impact. However, there are many things we can't know in advance of a nuclear war that might mitigate some damage or make it much worse. The unknown unknowns make a definitive conclusion about the possible extinction of humans impossible.

Would it be possible to survive in countries that were attacked? Yes, if you had a fallout shelter that was deep enough or far enough from the immediate effects of blast and fires, and if you stayed in it long enough so that you could safely come out (a week or more considering the radioactivity and possibly mainly at night to avoid ultraviolet radiation) and if you had enough supplies to last you for several years until you could produce more food, and if you did not go crazy from being cooped up in a hole in the ground knowing about all the deaths, and if you had enough weapons to keep out all the other hungry people, then you could survive.

While it is important to point out the consequences of nuclear winter, it is also important to point out what may not be the consequences. It is very difficult to predict the extinction or survival of species due to the interlocking nature of ecosystems. For example, there would likely be extreme health problems facing humans, including the spread of infectious diseases that could decimate surviving populations. Nevertheless, although extinction of our species was not ruled out in initial studies of nuclear winter by biologists, it now seems based on current work that this would be unlikely.

As discussed in Chapter 13, in Australia, New Zealand, and most of the Southern Hemisphere, large numbers of humans would likely survive at least the first few years after the war. Also, Earth will not be plunged into an ice age. Ice sheets, which covered North America and Europe only 18,000 years ago and were more than 3 km (2 miles) thick, take many thousands of years to build up from annual snow layers, and the climatic disruptions from a nuclear war would not last long enough to produce them. The oxygen consumption by the fires would be inconsequential, as would the effect on the atmospheric greenhouse by carbon dioxide production. The consequences of nuclear winter would be extreme enough without these additional effects, however.

While the extinction of humans might not be likely, a glance at Figure 13.5 indicates that only a tiny fraction of the populations of Canada, the United States, Europe, Russia, or China would survive a war between the United States, Russia, and their allies. Of course, much of their infrastructure would also be destroyed, and many more will perish after the second year than was considered in Figure 13.5. In contrast, much of the population of the Global South and of the tropics would survive and would have suffered little loss of infrastructure. Hence, it is likely that new countries and new cultures would dominate the post-war world. Those most likely to fight a nuclear war are those most likely to die from it. Hopefully, the new world powers would be smart enough to not repeat the mistakes of the current powers.

No matter who starts a nuclear war, the northern countries in the Northern Hemisphere that now have the coldest temperatures and shortest growing seasons are the most vulnerable to effects of a cooling climate due to smoke from burning cities. Therefore, as discussed in the next chapter, it would make sense for the governments of those countries to reduce nuclear arsenals and not to grow, maintain, and modernize them as they currently are doing.

Chapter 15
Can We Avoid Nuclear War?

15.1 Deterrence

Many have argued that the only reason no global wars occurred during the 80 years following the end of World War II is the deterrence provided by nuclear weapons. Others think we have just been very lucky to have avoided a second nuclear war given the large number of near misses in Table 8.1. Nuclear states claim they maintain their arsenals not to use them but for deterrence. This means that they threaten to use them if attacked, hoping that the threat will prevent any attacks using nuclear weapons. The "logic" of deterrence is flawed, as has been pointed out by former Secretaries of State George Shultz and Henry Kissinger, former Secretary of Defense William Perry, and Senator Sam Nunn (2007, 2008, 2010, 2011). Here, we present several reasons why persisting in the belief that deterrence justifies nuclear arsenals is flawed reasoning.

First, as we have seen in Table 8.1, nuclear weapons have accidentally come close to use several times, despite the arsenals being meant for deterrence. In fact, such accidental use is only possible because the weapons exist in the first place.

Second, nuclear weapons are not likely to deter attacks by terrorists or by cyberwarfare. Such attacks continue against nuclear nations. Those perpetuating them cannot be attacked with nuclear weapons, even if they could be identified; thus, they do not care about any claims of deterrence. Some terrorists may be state-sponsored, and cybercriminals may be state actors and not just individuals. There may have been cases where such behavior was deterred when the prospect of discovery of the state action was deemed too hard to hide. However, non-state actors are hard to target with nuclear weapons and are unlikely to be deterred. In fact, they may even be emboldened.

Third, nuclear weapons have not deterred multiple attacks on nuclear nations with conventional weapons. The Soviet Union took over Eastern Europe after World War II, and the only nuclear arsenal at the time—that of the United States—did not deter them. Nuclear-armed Israel was attacked in the Yom Kippur War by Egypt, Syria, and a coalition of Arab States in October 1973, and by Hamas in October 2023, undeterred by Israel's nuclear weapons. Argentina attacked the nuclear-armed United Kingdom to start the Falkland Islands (Malvinas) War in April 1982. Afghanistan defeated both the nuclear-armed Soviet

Earth in Flames. Owen Brian Toon and Alan Robock, Oxford University Press. © Oxford University Press (2025). DOI: 10.1093/9780197799734.003.0015

Union and nuclear-armed United States. The nuclear-armed United States lost the Vietnam War. At the time of this writing, Ukraine is fighting a tough battle against an invading Russian army. Despite Russian threats to use nuclear weapons, Ukraine and its allies have not been deterred from fighting this Russian aggression. It may be argued that nuclear weapons have deterred NATO from directly fighting Russia.

Fourth, do nuclear weapons deter attacks by other nuclear states? The militaries of India and China have fought recently along their border in the Himalayas, causing dozens of deaths for the first time in decades (Chervin 2024). Although they strictly followed rules on both sides that prohibited the use of guns and instead hit each other with sticks and stones, this reminds us that the two most populous nations on Earth maintain their nuclear arsenals to deter conflict. One could argue, though, that the magnitude of the conflict was lessened by the threat of nuclear conflict. In 2022, Vladimir Putin tried to use the threat of nuclear war to stop other countries from interfering in Russia's invasion of Ukraine. However, those other states, several of them with nuclear weapons of their own, were not deterred from providing military support to Ukraine. They may, however, have been deterred from directly attacking Russia. In 1998, India and Pakistan tested nuclear weapons. In 1999, Pakistan launched a conventional attack on part of Indian-occupied Kashmir, apparently assuming its nuclear arsenal would prevent India from attacking Pakistan. India did respond with conventional forces, and Pakistan was eventually forced to withdraw. It is impossible to know if nuclear deterrence works between nuclear states. The real question is, even if you believe that deterrence has worked, will deterrence last forever? It clearly is an unstable situation, as nuclear war can start unintentionally (see Table 8.1).

Fifth, as we explained in our 2012 article no country needs nuclear weapons to deter a nuclear attack because of "self-assured destruction" (Robock and Toon 2012). If one country conducts a first strike with their nuclear arsenal, they would produce so much climate change because of the smoke generated by their attack that agriculture would fail and everyone in their own country would suffer along with the rest of the world. High latitude countries, including Russia, China, and the United States, are highly susceptible to self-assured destruction. If there were enough smoke, which certainly either the United States or Russia could produce on their own, there would likely be nuclear winter, and most people in that attacking country, as well as the rest of the world, would perish. The realization of the climatic impacts of nuclear war should self-deter nations.

Since deterrence only works if countries believe other countries are insane, and would kill themselves if they initiated an attack, is there a rational reason that remains for keeping nuclear arsenals? We have been lucky for the past

80 years that nuclear weapons have not been used again, despite many close calls. However, this does not mean that we can safely keep them. Moreover, many credible scenarios could result in nuclear war should things go awry.

It may be rational, however, to maintain a small nuclear arsenal that is not capable of producing global famine. When we ask an audience how many nuclear weapons a country needs to threaten to use against the capital of an enemy in order to deter any attack from that enemy, the answer is usually "one." If a nuclear nation maintained a nuclear arsenal of 10 nuclear weapons, and they were invulnerable, perhaps based on a nuclear submarine, wouldn't that be enough of a deterrence? Only the current North Korean arsenal of perhaps 50 weapons may fall in that category. All the other nuclear nations have the capacity to produce catastrophic indirect climate change that would come back to affect them, even if no bombs were dropped on them.

15.2 How Do We Avoid a Nuclear War?

As discussed in previous chapters, a nuclear war would be a catastrophe for humanity, a catastrophe caused by humans rather than by natural events such as a supervolcano eruption or an asteroid impact. A nuclear war could occur by accident, be initiated intentionally by malevolent actors, or arise as an unintended consequence of conflict started as a low-level, local dispute. The question is: how can we avoid the same fate as the dinosaurs?

Unfortunately, we do not live in utopia. All wars are not going to stop, and no set of world leaders is likely to suddenly get rid of all nuclear weapons. Fortunately, we had been in a long period of building down nuclear arsenals, as illustrated in Figure 7.1. Every US president since Ronald Reagan and every Russian leader since Mikhail Gorbachev reduced the number of weapons through a series of treaties. Unfortunately, reductions since 2010 have been primarily due to disposal of old weapons planned from previous time periods. Treaties limiting weapons have either been dismantled or put on hold. In addition, the other nuclear powers have not reduced their arsenals significantly and have not engaged in making treaties to do so. Currently, given the war in Ukraine, which could escalate to nuclear conflict, it seems difficult to imagine new treaties in the near future. However, Russian threats to use nuclear weapons against a non-nuclear country may reawaken humanity to the threats of nuclear warfare and eventually lead to new treaties and weapons reductions.

Preventing nuclear war will likely come in two stages. The first stage is to make it much harder to use the existing nuclear arsenals and to stop proliferation. The second stage is to dismantle the nuclear arsenals; after all, if nuclear weapons exist, they can be used.

15.2.1 Making It Harder to Use Existing Weapons

Several steps could be taken immediately to lessen the chance that US nuclear weapons would be used. The United States has about 400 nuclear-armed missiles in underground silos in Colorado, Wyoming, Montana, Nebraska, and North Dakota. The Russians know where they are, as does anyone using the internet. An attack would destroy the missiles, so the United States has a policy of keeping them on hair-trigger alert and an option to launch them on warning of an attack, on the order of the president. What if the warning is a mistake? There would only be 10 or 20 minutes at most between a warning and an order to launch, not nearly enough time for a considered, rational decision and verification that the attack is real. As discussed in Chapter 8, many such warnings have been in error (Jacobsen 2024). But the US arsenal also includes submarine missiles, which are presently invulnerable, as well as nuclear weapons that could be loaded on aircraft and put on runway alert, but can be recalled if needed. If the United States wants to use its weapons as a deterrent, submarines and aircraft are more than enough. So, taking the land-based missiles off hair-trigger alert and then dismantling them would be a major step toward reducing the threat of nuclear war. Or the United States could simply leave most of the silos empty, which would present the same number of targets to an enemy but prevent the need to launch them quickly. This would also send a powerful signal to non-nuclear nations that we are serious about nuclear arms reductions. This action would reverse the current US plan to spend more than $200 billion to build new Sentinel missiles to replace the current ones.

Arguments against getting rid of this leg of our nuclear triad (land-, sea-, and air-based missiles) include that it will make us look weak. But the United Kingdom, for example, only has sea-based nuclear weapons, and that is sufficient for them. Another argument is that it will reduce the nuclear parity with Russia. However, when the Soviet Union was collapsing in the 1980s, President George H. W. Bush unilaterally halved the US-deployed nuclear arsenal, as a means of lessening international tension, without any treaty requiring him to do so, and later the Russians followed suit. This is an action the current US president could do today. There would certainly be pressure from the US military, particularly the Strategic Command, which oversees the land-based missiles, and from military contractors who benefit from this branch of the military-industrial complex. As President Eisenhower warned us on January 17, 1961, in his farewell speech, "In the councils of government, we must guard against the acquisition of unwarranted influence, whether sought or unsought, by the military-industrial complex. The potential for the disastrous rise of misplaced power exists, and will persist." A strong president has the authority to take this step, but it would not be easy because Congress has embedded in law bans against reducing US intercontinental ballistic missiles (ICBMs).

Another step is to change the declared US policy and announce that we will not be the first to use nuclear weapons. Now we often hear that "all options are on the table." This is a threat to use nuclear weapons in response to non-nuclear aggression. But we have more than sufficient "conventional" weapons to respond if we think that violence is the correct response. Hopefully, an effective leader would use negotiations to prevent us getting to a situation where nuclear weapons might be used. But an important declaration for the United States to lessen the possibility of the use of nuclear weapons is one of "no first use." Currently, India and China have a declared no-first-use policy. They recognize that any use of nuclear weapons could lead to escalation and a global nuclear holocaust. If the United States were to join them, it would be a major step toward a safer world.

President Harry Truman was the only person on Earth to ever order the use of nuclear weapons in war. At the time, the tremendous power of what today would be considered a small atomic bomb was shocking. Truman decided that the future US policy would be that only the president could authorize the use of nuclear weapons; an aide always follows the president with a satchel, called "the football," with the nuclear launch codes. This means that one person can destroy the world, with no way to delay or think about the process if everyone follows orders. In Russia, it is believed, that three people must agree before nuclear weapons can be launched. What if we were to change the rules in the United States so that the order to use nuclear weapons would have to be made by a small committee of political and military figures, including the vice president and the Speaker of the House, so that it could be done deliberately? A strong argument for this is made by Perry and Collina (2019).

Unfortunately, as Ellsberg (2017) so eloquently explained, the "football" may not be the only means to launch a nuclear weapon, because many lower-level military officers may have the ability to launch nuclear weapons if they perceive a threat to the United States and are unable to communicate with the president. When Alan mentioned this ability of lower-level officers to launch nuclear weapons to retired General C. Robert Kehler, former commander of United States Strategic Command, at a workshop at Stanford University in February 2017, he was told that it was not correct because "there is always a president." In theory, General Kehler is correct, since if the president and vice president, for example, are killed in an attack, the Constitution and the Congressional Presidential Succession Act provide for a line of succession. In the aftermath of a nuclear or conventional attack, however, it may not be possible to quickly communicate with whomever is president, or to even know who the president is, as described in Annie Jacobsen's book. Uncertainty about who can launch nuclear weapons is another argument for nuclear abolition.

15.1.2 Non-Nuclear Zones

Only nine nations have nuclear weapons. Why have the other 186 nations chosen not to make nuclear weapons? How to make nuclear weapons is no longer a secret. All that is needed is plutonium or highly enriched uranium and a moderate engineering effort. It is true that plutonium and the means to enrich uranium are somewhat hard to come by, but several nations have them, and centrifuge technology may be simple to develop and easy to hide (Kemp 2014). At least one non-nuclear nation, Iran, has an active program to enrich uranium and did have a nuclear weapon design program. Japan and South Korea, uncertain of US support and wanting to counter North Korea, have recently considered nuclear weapons, as have Saudi Arabia and Egypt because of their concern about Iran and Israel.

Non-nuclear nations come in two varieties. One set, while not having nuclear weapons of their own, is allied with nuclear nations and counts on them to use their "nuclear umbrella" as a protection. Remarkably, the only nation to have suffered atomic bombing, Japan, is one of these. The North Atlantic Treaty Organization (NATO) includes three nuclear states—the United States, the United Kingdom, and France—and their Article 5 requires other member states to come to the defense of members that are attacked. Curiously, Article 5 has only been invoked once in the history of NATO, after the September 11, 2001, terrorist attacks on the United States. You would think that this would make other nuclear states question the value of nuclear deterrence, since the only Article 5 attack (by a non-governmental group) was not deterred by the large NATO nuclear arsenal. And some of the other NATO members allow nuclear weapons on their territory, which may make them less safe if there were a war.

The other variety of non-nuclear states, which is the vast majority of states, has decided that nuclear weapons make them less safe. In fact, there are no nuclear weapons in Latin America or Africa or in several parts of Asia. See Figure 7.6 for a map of the world that shows the nuclear status of each nation. Figure 7.5 lists treaties that ban nuclear weapons in much of the world. Currently about one-third of the world's population lives in regions that the United Nations recognizes as nuclear-free zones. The 1969 Treaty for the Prohibition of Nuclear Weapons in Latin America and the Caribbean, sometimes called the Treaty of Tlatelolco, earned the 1982 Nobel Peace Prize for Alfonso García Robles of Mexico. Brazil and Argentina both began programs to develop nuclear weapons but then canceled them.

Some current non-nuclear countries once had nuclear weapons or nuclear weapons programs. South Africa developed nuclear weapons with the assistance of Israel. Israel and South Africa may have tested one in the atmosphere, though attempts at underground testing were abandoned under pressure from the Soviet Union, the United States, France, and other governments. After constructing six nuclear weapons, South Africa decided to cancel their nuclear program and dismantle their weapons in the 1990s. South Africa is the only country to have developed nuclear weapons and then discarded them. However, as illustrated in Figure 7.3, several former Soviet republics—Ukraine, Kazakhstan, and Belarus—returned nuclear weapons to Russia when they left the Soviet Union.

Many other countries once had nuclear programs but abandoned their nuclear weapons programs when the Nuclear Non-Proliferation Treaty came into force in 1970 or when the Cold War ended. Brazil and Japan still produce fissionable materials. Many countries could develop nuclear weapons if they wanted to invest the resources but have decided that the money could be better spent elsewhere and that not having nuclear weapons would remove them from the target lists of those nations that have them. When the rest of the world learns to make these non-nuclear countries their role models, the world will move to a safer state.

15.1.3 ICAN and the Nobel Peace Prize Pressure Politicians to Eliminate Nuclear Weapons

Alfred Nobel's father ran an armaments factory for the Russian tsar in St. Petersburg in the mid-nineteenth century, so it was natural that Alfred developed an interest in explosives. Nitroglycerin, invented by the Italian Ascanio Sobrero, was very unstable and dangerous to work with, but Alfred Nobel figured out that by mixing it with clay he could produce a safer way to handle it. This invention of dynamite made him one of the richest men on Earth. Dynamite was used widely for industrial purposes, such as mining and building demolition, but also for war. In a story told by Al Gore in his Nobel Prize acceptance speech, a premature obituary of Alfred Nobel labeled him "The Merchant of Death" because of his invention of dynamite. Nobel then decided he should work for peace. Seven years later, Nobel devoted part of his fortune to endow an annual Peace Prize.

The Nobel Peace Prize is one of the highest honors one can receive on Earth. And a number of them have gone to those seeking to lessen the dangers of nuclear war (Table 15.1). The penultimate such award so far went to the International Campaign to Abolish Nuclear Weapons (ICAN) in 2017. Beatrice Fihn, the director ICAN, said this as she accepted the Nobel Peace Prize:

If only a small fraction of today's nuclear weapons were used, soot and smoke from the firestorms would loft high into the atmosphere—cooling, darkening and drying the Earth's surface for more than a decade. It would obliterate food crops, putting billions at risk of starvation. Yet we continue to live in denial of this existential threat. . . . The story of nuclear weapons will have an ending, and it is up to us what that ending will be. Will it be the end of nuclear weapons, or will it be the end of us? One of these things will happen. The only rational course of action is to cease living under the conditions where our mutual destruction is only one impulsive tantrum away.

Table 15.1 Nobel Peace Prizes for Advocating Nuclear Disarmament (from https://www.nobelprize.org/prizes/lists/all-nobel-peace-prizes)

Year	Nobel Laureates
1959	**Philip Noel-Baker** "He engaged in intense efforts to prevent nuclear war between the United States and the Soviet Union."
1962	**Linus Pauling** "He spoke and wrote against the nuclear arms race, and he was a driving force in the Pugwash movement. . . . He was one of the prime movers who urged the nuclear powers the USA, the Soviet Union and Great Britain to conclude a nuclear test ban treaty."
1982	**Alva Myrdal** "She worked actively to persuade the superpowers to disarm. The nuclear race was a major concern, and she fought for nuclear weapons-free zones in Europe." *and* **Alfonso García Robles** "played a key part in the laborious efforts to make Latin America a nuclear-free zone. . . . He was lauded as 'Mr. Disarmament.'"
1985	**International Physicians for the Prevention of Nuclear War** "IPPNW held annual congresses to tell the world about the consequences of nuclear war. Extensive nuclear explosions could prevent sunlight from reaching the earth. The resulting drop in temperature would cause a 'nuclear winter.' The organization recommended a nuclear test ban and demanded that the great powers should refrain from first use in conflict situations."
1995	**Joseph Rotblat** and **Pugwash Conferences on Science and World Affairs** "for their efforts to diminish the part played by nuclear arms in international politics and, in the longer run, to eliminate such arms."
2005	**International Atomic Energy Agency (IAEA)** and **Mohamed ElBaradei** "for their efforts to prevent nuclear energy from being used for military purposes and to ensure that nuclear energy for peaceful purposes is used in the safest possible way."
2009	**Barack Obama** "for his extraordinary efforts to strengthen international diplomacy and cooperation between peoples."
2017	**International Campaign to Abolish Nuclear Weapons (ICAN)** "for its work to draw attention to the catastrophic humanitarian consequences of any use of nuclear weapons and for its ground-breaking efforts to achieve a treaty-based prohibition of such weapons."
2024	**Nihon Hidankyo** (the Japan Confederation of A- and H-Bomb Sufferers Organizations) "for its efforts to achieve a world free of nuclear weapons and for demonstrating through witness testimony that nuclear weapons must never be used again."

Half of the 2017 ICAN Nobel Peace Prize was awarded "for its work to draw attention to the catastrophic humanitarian consequences of any use of nuclear weapons." ICAN took part in organizing three international conferences on the Humanitarian Impact of Nuclear Weapons in Oslo, Norway, March 4–5, 2013; in Nayarit, Mexico, February 13–14, 2014; and in Vienna, Austria, December 8–9, 2014. They were attended by representatives of more than 100 nations, who were educated and convinced that the humanitarian impacts of nuclear weapons required their abolition. The humanitarian impacts that were discussed were twofold. One was the obvious direct impacts on humans where bombs would be dropped. *Hibakusha* (survivors) of the Hiroshima and Nagasaki bombings gave heartbreaking testimony of the horrors. The other potential humanitarian impacts would be from the indirect effects of climate change and famine, for which we use the term "nuclear winter." At the Oslo conference, Ira Helfand presented our work on the agricultural impacts of a war between India and Pakistan. In Nayarit, Alan presented an update and emphasized that a full nuclear winter is still possible with current arsenals from a war between the United States and Russia. In Vienna, Michael Mills presented our results about ozone loss following a nuclear war. Thus, our work contributed to the message of ICAN and to the impetus for a UN treaty to ban nuclear weapons.

The second half of the 2017 ICAN Nobel Peace Prize was awarded "for its ground-breaking efforts to achieve a treaty-based prohibition of such weapons." On July 7, 2017, 122 of the United Nations member states adopted the Treaty on the Prohibition of Nuclear Weapons (TPNW). On January 22, 2021, 90 days after the treaty had been ratified by 50 states, it came into force. So far, none of the nine nuclear weapons states has ratified it. Once the United States, Russia, France, the United Kingdom, China, India, Pakistan, Israel, and North Korea stop pretending that nuclear weapons bring them security and accept that the use of them would kill most people in their own countries because of the impacts on climate and food, and thus they would be acting as suicide bombers, the world can dedicate its efforts toward solving other important problems, including global warming.

ICAN's Nobel Peace Prize is not the only one recognizing nuclear winter. In his acceptance speech for the 2007 Nobel Peace Prize for his work on the greenhouse effect and global warming, Al Gore noted:

Indeed, without realizing it, we have begun to wage war on the Earth itself. Now, we and the Earth's climate are locked in a relationship familiar to war planners: "mutually assured destruction." More than two decades ago, scientists calculated that nuclear war could throw so much debris and smoke into the air that it would block life-giving sunlight from our atmosphere, causing a "nuclear winter." Their eloquent warnings here in Oslo helped galvanize the world's resolve to halt the nuclear arms race.

Now science is warning us that if we do not quickly reduce the global warming pollution that is trapping so much of the heat our planet normally radiates back out of the atmosphere, we are in danger of creating a permanent "carbon summer."

As the American poet Robert Frost wrote, "Some say the world will end in fire; some say in ice." Either, he notes, "would suffice."

But neither need be our fate. It is time to make peace with the planet.

In his 1995 Nobel Prize in Chemistry speech, Paul Crutzen noted:

Although I do not count the "nuclear winter" idea among my greatest scientific achievements (in fact, the hypothesis cannot be tested without performing the "experiment," which it wants to prevent), I am convinced that, from a political point of view, it is by far the most important, because it magnifies and highlights the dangers of a nuclear war and convinces me that in the long run mankind can only escape such horrific consequences if nuclear weapons are totally abolished by international agreement. I thus wholeheartedly agree in this respect with Joseph Rotblat and the Pugwash organization, this year's recipients of the Nobel Prize for Peace.

In his 1985 Nobel Peace Prize lecture on behalf of the International Physicians for the Prevention of Nuclear War, Dr. Yevgeny Chazov of the Soviet Union noted:

Physicians have demonstrated to the whole world that not only would nuclear war spell the end of civilization, it would also prejudice the existence of life on Earth. My conscience, and I am sure the same applies to many of my colleagues in IPPNW, was staggered primarily by the total number of victims in nuclear war. The human mind finds it difficult to comprehend the figure of 2,000 million victims. As they say, one death is death, but a million deaths are statistics. For us, physicians, life is the aim of our work and each death is a tragedy. As people constantly involved in the care of patients, we felt the urge to warn governments and peoples that the critical point has been passed: medicine will be unable to render even minimal assistance to the victims of a nuclear conflict—the wounded, the burned, the sick—including the population of the country which unleashes nuclear war. Even rough estimates show it would require efforts of at least 30 million physicians, 100 million nurses and technical personnel. These, of course, are absolutely unrealistic figures. In the world today there are around 3.5 million physicians and about 7.5 million nurses. Treatment of a few hundred patients suffering from burns as a result of a major fire can rapidly exhaust the burn cure resources of a large city. Where, then, can the resources be found to treat thousands and millions of casualties? Physicians and hospitals will face an insoluble problem, even if we discount the appalling conditions of "nuclear winter" which is bound to cap the catastrophe. Besides, in a nuclear war many physicians and nurses will be killed and many hospitals destroyed.

US President Barack Obama's 2009 Nobel Peace Prize lecture noted:

One urgent example is the effort to prevent the spread of nuclear weapons, and to seek a world without them. In the middle of the last century, nations agreed to be bound by a treaty whose bargain is clear: All will have access to peaceful nuclear power; those without nuclear weapons will forsake them; and those with nuclear weapons will work towards disarmament. I am committed to upholding this treaty. It is a centerpiece of my foreign policy. And I'm working with President Medvedev to reduce America and Russia's nuclear stockpiles.

But it is also incumbent upon all of us to insist that nations like Iran and North Korea do not game the system. Those who claim to respect international law cannot avert their eyes when those laws are flouted. Those who care for their own security cannot ignore the danger of an arms race in the Middle East or East Asia. Those who seek peace cannot stand idly by as nations arm themselves for nuclear war.

15.2 Nuclear Weapons Treaties

Since the 1960s, a series of treaties has defined nuclear free zones and sought to control the numbers of nuclear weapons in the United States and Russia. Figure 7.5 lists all the international treaties banning nuclear weapons in certain places. Table 15.2 lists the treaties that limit the size of US and Russian arsenals. Table 15.3 lists treaties that prohibit testing of nuclear weapons. The testing treaties have been instrumental in stopping dangerous bomb testing in the atmosphere, and then preventing any testing, even underground. The most recent treaty, the 1996 Comprehensive Test Ban Treaty, has been ratified by 148 states, but it will not come into force until the following countries ratify it: China, North Korea, Egypt, India, Iran, Israel, Pakistan, and the United States. While this list includes six of the current nuclear states, and one (Iran) that appears to be trying to develop a nuclear option, the treaty has nevertheless served as strong taboo against continued testing. Russia has not tested any weapons by exploding them since 1990, and the United States has not since 1992. The United Kingdom last tested in 1992 at the US test site, France and China in 1996, and India and Pakistan in 1998. The only nation that continues to test nuclear weapons in the twenty-first century is North Korea, which has conducted six underground tests,

Table 15.2 Treaties to Limit the Number of Weapons (from https://www.armscontrol.org/treaties and https://en.wikipedia.org/wiki/Nuclear_disarmament)

Year Signed	Treaty
1968 Into force 1970 Extended indefinitely in 1995	**Nuclear Non-Proliferation Treaty (NPT):** An international treaty (currently with 191 member states) to limit the spread of nuclear weapons. The treaty has three main pillars: nonproliferation, disarmament, and the right to peacefully use nuclear technology. But Article 6, which commits parties to "pursue negotiations in good faith on effective measures relating to cessation of the nuclear arms race at an early date and to nuclear disarmament, and on a treaty on general and complete disarmament," has been ignored.
1972 U.S. withdrew 2002	**Interim Agreement on Offensive Arms (Strategic Arms Limitation Treaty—SALT I):** The Soviet Union and the United States agreed to a freeze on the number of ICBMs and SLBMs that they would deploy.
1972 U.S. withdrew 2002	**Anti-Ballistic Missile Treaty (ABM):** The United States and Soviet Union could deploy ABM interceptors at two sites, each with up to 100 ground-based launchers for ABM interceptor missiles. In a 1974 Protocol, the United States and the Soviet Union agreed to only deploy an ABM system to one site so that they could not be used as a nationwide defense.
1979	**Strategic Arms Limitation Treaty (SALT II):** Replacing SALT I, SALT II limited both the Soviet Union and the United States to an equal number of ICBM launchers, SLBM launchers, and heavy bombers. The treaty also placed limits on multiple independent reentry vehicles (MIRVs).
1987 U.S. withdrew 2019	**Intermediate-Range Nuclear Forces Treaty (INF):** Created a ban on ground-launched and cruise missiles with ranges from 500 to 5000 km and created an intrusive verification regime for the United States and Soviet Union.
1991 Ratified 1994	**Strategic Arms Reduction Treaty (START I):** Limited long-range nuclear forces in the United States and the newly independent states of the former Soviet Union to 6000 attributed warheads on 1600 ballistic missiles and bombers.
1992 U.S. withdrew 2020	**Open Skies Treaty:** Established a regime of unarmed aerial observation flights over state territories and enhances mutual understanding of and increase transparency in military forces and activities.
1993 Never put into force	**Strategic Arms Reduction Treaty II (START II):** START II was a bilateral agreement between the United States and Russia that attempted to commit each side to deploy no more than 3000 to 3500 warheads by December 2007 and included a prohibition against deploying MIRVs on ICBMs.

2002 Into force 2003	**Strategic Offensive Reductions Treaty (SORT or Moscow Treaty):** This very loose treaty is often criticized by arms control advocates for its ambiguity and lack of depth. Russia and the United States agreed to reduce their "strategic nuclear warheads" (a term that remained undefined in the treaty) to between 1700 and 2200 by 2012. Was superseded by New START Treaty in 2010.
2010 Into force 2011 Russia suspended in 2023, but said they would honor provisions	**New Strategic Arms Reduction Treaty (New START):** Replaces SORT treaty, reduces deployed nuclear warheads by about half, and will possibly remain in force until 2026.

Table 15.3 Treaties That Limit the Testing of Weapons (from https://www.armscontrol.org/treaties and https://en.wikipedia.org/wiki/Nuclear_disarmament)

Year Signed	Treaty
1963	**Partial Test Ban Treaty:** Prohibited all testing of nuclear weapons except underground.
1974	**Threshold Test Ban Treaty (TTBT):** This treaty between the United States and the Soviet Union established a nuclear threshold through the prohibition of the testing of new or existing nuclear weapons with a yield exceeding 150 kilotons.
1976	**Peaceful Nuclear Explosions Treaty (PNET):** This treaty between the United States and the Soviet Union prohibits peaceful nuclear explosions not covered by the Threshold Test Ban Treaty and verifies all data exchanges and visits to sites of explosions through national technical means.
1996 not yet in force	**Comprehensive Test Ban Treaty (CTBT):** An international treaty (currently with 181 state signatures and 148 state ratifications) that bans all nuclear explosions in all environments. While the treaty is not in force, Russia has not tested a nuclear weapon since 1990, and the United States has not tested a nuclear weapon since 1992.

the last in 2017. The United States uses numerical simulation models to test new designs on some of the fastest computers in the world at the Lawrence Livermore National Laboratory and the Los Alamos National Laboratory. They set an example for the world, but some in the United States want to resume nuclear testing, not because of a need to validate or develop weapons but as a political statement.

The treaties that limit the number of nuclear weapons, apart from the Nuclear Non-Proliferation Treaty (NPT), have been between the countries with by far the two largest nuclear arsenals: the United States and the Soviet Union (now Russia). The treaties have led to a gradual reduction in their nuclear arsenals and are enforced by an intrusive on-site inspection regime through cooperation on both sides. The treaties have been imperfect, but effective. The current treaty, New START, limits the deployment on each side to 1550 strategic (long-range) nuclear delivery systems, but airplanes only count as carrying one warhead each, so the effective limit is about 2000 on each side. (As we have seen above, this is way more than needed to produce a true nuclear winter, with temperatures below freezing in the summer over continents.) New START also limits the number of deployed and non-deployed ICBM and submarine-launched ballistic missile (SLBM) launchers to 800 on each side. The New START treaty does not cover tactical nuclear weapons, those designed for the battlefield, nuclear mines, artillery, short- and medium-range ballistic missiles, cruise missiles, and gravity bombs to be dropped from airplanes or reserve strategic warheads in storage. The New START treaty originally covered 10 years, but in early 2021 it was extended for five years by agreement of the two sides. Unfortunately, Russia suspended the treaty in 2023, but said it will continue to honor its terms.

The 1968 NPT treaty is the bedrock treaty attempting to limit the spread of nuclear weapons and eventually eliminate them. It has had some success, but in other respects it has been a failure. The treaty makes a distinction between the five nuclear-weapons states at the time it was written (the United States, Russia, the United Kingdom, France, and China) and the rest of the world, somehow implying that they had a right to nuclear weapons and the rest of the world did not. This has been an impediment to the elimination of nuclear weapons. Why should a country such as Iran listen to these five countries telling it not to become nuclear while they reserve the right to keep their nuclear weapons? It is like someone sitting on a barstool telling others not to drink. It is, "Do as I say, not as I do." In fact, Article 6 of the NPT says that "each of the Parties to the Treaty undertakes to pursue negotiations in good faith on effective measures relating to cessation of the nuclear arms race at an early date and to nuclear disarmament, and on a treaty on general and complete disarmament under strict and effective international control." In the 50 years since the treaty was ratified, this part has been an abject failure. In fact, now there are nine nuclear-weapons states (four not parties to the NPT), not five, and none of the five has moved toward disarmament. While the arms race ended in the 1980s, due to scientists explaining the dangers of nuclear winter and massive protests in the United States and United Kingdom, nuclear weapons have spread in parallel with the spread of nuclear power, which legitimizes acquiring the ability to produce the nuclear fuel needed for nuclear weapons. There are 32 nations with nuclear reactors and 53 with research reactors. The NPT has review conferences every five years to try to make progress, and annual meetings through the Conference on Disarmament in Geneva, but they have not been at all effective in the last decade or two.

While six of the nine nuclear nations are in Asia, there are many locations on Earth, and even beyond Earth, where nuclear weapons are banned by treaty (Figures 4.4 and 4.5, Table 14.1). This includes Antarctica, Latin America, Africa, the South Pacific, the seabed, and outer space. The most recent treaty, the TPNW, bans nuclear weapons everywhere. Unlike most of the other treaties, it was not negotiated by diplomats from nuclear states but was passed in the United Nations by non-nuclear states frightened by the prospect of the use of nuclear weapons and the indirect effects on those states even if no bombs were dropped on them. ICAN won the 2017 Nobel Peace Prize for warning the world about the humanitarian consequences and for getting the treaty passed in the General Assembly of the United Nations, which was not subject to veto in the United Nations Security Council. It is not a coincidence that the five permanent members of the Security Council, those with veto power, are the first five nations to acquire nuclear weapons.

The 2017 Nobel Peace Prize included an award of about US$1,060,000. ICAN has been using this money ever since in a campaign to get nations to sign and ratify the TPNW. As discussed above, the numbers that have ratified the treaty keep growing, but the nuclear states continue to ignore the will of the rest of the world. With other treaties, education and time have resulted in either ratification and enforcement of the treaty, (or at least de facto compliance with the treaty provisions). With the TPNW, at the time of this writing, it is still an ongoing process.

15.3 What Can You Do?

The threat described in this book is pretty depressing and overwhelming. So, what do you do with this information? The most natural reaction is to try to forget it. As Mark Twain is alleged to have said, "Denial ain't just a river in Egypt." But if you, like us, want to try to rid the world of the threat of nuclear weapons, there are several things you can do. At the time of this writing, the wars in Ukraine and Gaza, getting your kids a good education, climate change, health care, and the economy are the overwhelming short-term concerns. But if we want a future for us and our kids and grandkids after these and other short-term issues pass, we must devote some effort to ridding the world of the nuclear pandemic.

15.3.1 Education

First, people have to understand the threat. We have tried to get the word out, first in scientific articles, which are peer-reviewed and published in the top scientific journals, including *Science, Nature,* and *Proceedings of the National Academy of Sciences of the United States.* We have also written articles that should be more accessible to the non-specialist, including in *Scientific American, Bulletin of the Atomic Scientists,* and *Huffington Post.* We have written encyclopedia articles, used Twitter (now X), and given TED talks and lectures to community groups and the public. We include this information in courses we teach at the University of Colorado and Rutgers University. And we have written this book. A complete list of our publications and our efforts is at http://climate.envsci.rutgers.edu/nuclear/. Educate yourself. Educate your friends and neighbors.

15.3.2 Take Personal Political Action

In a democracy, elected members such as Congress and the president in the United States take actions that help their constituents and improve their chances for reelection. But to do this, they need to know what their constituents want. So, if you live in the United States, take an hour and write a letter to your representative and two senators in Congress and to the president. Or if you live in another open democracy, such as the United Kingdom, France, Israel, India, or Pakistan, contact your political leaders. Tell them of your concern about the threat of nuclear weapons and your desire that they quickly reduce nuclear arsenals and work with other nations of the world to abide by Article 6 of the NPT toward the goal of a world free of nuclear weapons. Many countries are wasting money building expensive new weapons that can never be used; let's hope they will just sit in the ground and rot. In the United States, ask your government to cancel plans to spend more than a trillion dollars to modernize weapons that cannot be used without killing us all. In India and Pakistan, ask your governments to stop enlarging their nuclear arsenals. In other nuclear-armed countries such as China, North Korea, or Russia, it may be more difficult to write to the political representatives in your area. In that case, do what is possible to educate people about the dangers of nuclear weapons. In China, suggest that your leaders halt the construction of new missiles and warheads and spend the money instead on expanding your economy.

15.3.3 Take Organized Political Action

Several organizations work together to advocate for nuclear-threat reduction. By joining and supporting them, your voice will be multiplied and heard. In the 1980s, these organizations—such as the National Committee for a Sane Nuclear Policy (SANE) in the United States and the Campaign for Nuclear Disarmament (inventor of the peace sign, ☮) in the United Kingdom—mobilized mass marches that got the attention of governments and were a major force in ending the nuclear arms race. Today, Global Zero, the Ploughshares Fund, and Sōka Gakkai, a peace-orientated Buddhist organization, are examples of such general organizations. ICAN is a coalition of more than 650 nongovernmental organizations in more than 100 countries promoting adherence to and implementation of the United Nations Treaty on the Prohibition of Nuclear Weapons. Find an organization in your country to join at https://www.icanw.org/partners. There are currently more than eighty different organizations in the United States; you can find them listed at https://www.icanw.org/united_states#countries-list. Join a local chapter and work together. A new organization, Students for Nuclear Disarmament (https://www.students4disarmament.org/), has also recently been formed. If you are young, get involved with them. It's your world to save.

There are also professional organizations such as the International Physicians for the Prevention of Nuclear War (winner of the 1995 Nobel Peace Prize) and its US affiliate, Physicians for Social Responsibility, and the Physicists' Coalition for Nuclear Threat Reduction. The Physicists' Coalition is hosted by the Arms Control Association and lobbies in Washington on nuclear issues. It has a project team of 14 lecturers, including Alan Robock, who present nuclear issues at Physics Departments around the United States, and have now recruited more than 1000 members. It successfully lobbied in 2020 for no nuclear testing and an extension of the New START Treaty. More recently, it advocated no first use of nuclear weapons, stopping the US modernization of the nuclear arsenal, and other policies.

Members of the Physicists' Coalition worked with Congress to add a requirement in the Fiscal Year 2021 U.S. National Defense Authorization Act for an "independent study on potential environmental effects of nuclear war (sec. 3171)" to "require the Administrator of the National Nuclear Security Administration to enter into an agreement with the National Academies of Sciences, Engineering, and Medicine to conduct a study on the non-fallout atmospheric effects of nuclear explosions. The study would assess the strengths and weaknesses of existing models in the areas of fire effects, soot generation and transport, radioactivity, and the atmospheric transfer of gasses. The provision would require the National Academies to submit a report on the study to the Administrator and the congressional defense committees no later than 18 months after the enactment of this Act. The provision would also require the Secretary of Defense to provide to the National Academies such information as necessary for the conduct of the study" and "would require the study to be conducted in consultation with the Secretary of Defense and the Director of National Intelligence." Although already past the deadline, at this writing, the study is under way. Alan and Brian gave invited presentations to the first meeting of this Independent Study on Potential Environmental Effects of Nuclear War Committee on February 23, 2023, but the study is not scheduled to be completed until sometime in 2025.

In an encouraging move, the United Nations General Assembly in December 2024 voted to launch a three-year study on "the physical effects and societal consequences of a nuclear war on a local, regional and planetary scale, including, inter alia, the climatic, environmental and radiological effects, and their impacts on public health, global socioeconomic systems, agriculture and ecosystems, in the days, weeks and decades following a nuclear war." They will convene a panel of 22 international experts to conduct the study, the first one they have done in more than 30 years.

Organized political action in the past and now has been effective at changing policies. Speaking truth to power sometimes works. The message of nuclear winter from us scientists, along with strong messages from the public, was instrumental in ending the nuclear arms race in the 1980s. But, despite substantial reductions in our nuclear arsenals, the problems of nuclear weapons and nuclear winter are still with us. We scientists are doing what we can to inform the public and our leaders of the problem. With help from the public, we are sure that together we can again be successful.

Glossary

Activity – The number of disintegrations per second of a radioactive material or the number of atoms that decay per second, usually measured in Becquerels (Bq) or Curies (Ci).

Airburst – A nuclear explosion detonated at an altitude—typically, thousands of feet—that maximizes blast damage. Because the fireball never touches the ground, an airburst produces less radioactive fallout than a groundburst.

Alpha particle – A particle emitted spontaneously from the nucleus of some radioactive elements, which is identical to the nucleus of a helium atom having two protons and two neutrons.

Atom – The smallest piece of an element that still retains the properties of the element. Atoms are electrically neutral. They have a central nucleus that contains most of the mass in the form of protons and neutrons and are surrounded by a cloud of electrons.

Atomic bomb – Usually refers to a nuclear weapon that is based on fission.

Assured destruction – The amount of destruction of the target done in the least favorable circumstances for creating damage.

Asteroid – A rocky object that probably formed inside the orbit of Jupiter where it was too warm in the early history of the solar system for ices to condense. Mercury, Venus, Earth, and Mars likely formed from the collisions of multitudes of asteroids. Sometimes called minor planets of the inner solar system.

Atomic mass unit (amu) – A unit used for the masses of atoms, molecules, and sub-atomic particles. It is defined to be one-twelfth of the mass of an atom of the most abundant isotope of carbon, ^{12}C.

Becquerel – A unit of activity equal to one disintegration per second, abbreviated Bq. Because of its small value, an EBq is often used, which is 10^{18} disintegrations per second. An older name for this a Curie, but in different units (1 Bq = 2.7×10^{-11} Curies).

Beta particle – An electron accompanied by an antineutrino in a process that converts a neutron to a proton, or a positron and a neutrino in a process that changes a proton to a neutron.

Blast wave – An abrupt jump in air pressure that propagates outward from a nuclear explosion, damaging or destroying whatever it encounters.

Billion – One thousand million.

Boosted fission weapon – A weapon in which extra neutrons are supplied to enhance the fission process. The extra neutrons are usually generated from the fusion reaction between deuterium and tritium that occurs as fission heats the material.

Comets – Objects primarily composed of water ice, but with small amounts of rock and other ices. Comets likely formed in the colder portion of the early solar system. Some asteroids may be remnants of comets in which the ice evaporated due to passing close to the Sun. There are billions of comets located beyond the orbit of Pluto in the Kuiper Belt or the Oort Cloud.

Conflagration – A moving fire driven by high winds.

Counterforce war – A war with targets designed to destroy the enemy's ability to conduct a military response.

Countervalue war – A war in which targets have great intrinsic value to the enemy such as population centers, cultural and historical locations, and economic centers.

Crater – A pit or depression in Earth's surface caused by an impact or a nuclear explosion.

Critical mass – The smallest mass of a fissionable material that will maintain a fission chain reaction. An explosion requires slightly more material than the critical mass, which is called a supercritical mass.

Cruise missile – A short-range unpiloted vehicle that that can evade radar by flying at low altitude and can be guided or automatically change course to locate its target with high accuracy.

Curie – An amount of radioactive material that produces 3.7×10^{13} nuclear disintegrations per second.

Cyclotron – A device that accelerates ions, such as electrically charged atoms, using an electric field and then uses a magnetic field to curve the path of the ions so that they circle back through the electric field and are accelerated further. The radius of the circle depends on the mass of the ion, so that ions such as ^{235}U and ^{238}U become separated slightly allowing uranium rich in the fissionable ^{235}U isotope to be collected.

Decay (or radioactive decay) – The decrease in activity of a radioactive material often due to the release of neutrons, alpha particles, or beta particles, sometimes accompanied by gamma radiation, a type of high-frequency light radiation.

Deuterium – An isotope of hydrogen having one proton, one neutron and one electron.

Direct radiation – Nuclear radiation produced in the first minute after detonation of a nuclear weapon, constituting the most immediate effect on the surrounding environment.

Dose – The amount of nuclear radiation absorbed per gram of absorbing material. It is usually measured in rads or Sieverts.

Dose equivalent – A measure of the biological effectiveness of a particular dose of radiation. It attempts to consider the types of tissue exposed, the type of radiation and other factors in a manner that equalizes the damage done. Usually measured in rems or Grays.

Dose rate – The rate at which a dose occurs in units such as rads per second or Sieverts per second.

Dynamic pressure – The air pressure that results from the wind behind the shock front of a blast wave.

Ecliptic plane – The plane of Earth's orbit around the Sun. The other solar system planets orbit the sun within about 7 degrees of the ecliptic plane.

Electromagnetic pulse (EMP) – An intense burst of radio frequency waves produced by a high-altitude nuclear explosion, capable of damaging electronic equipment over thousands of miles.

Electron – A negatively charged subatomic particle that is found outside the nucleus of atoms. Its mass is about 1836 times smaller than that of a positively charged proton.

Electron volt (ev) – An ev is the energy gained by an electron accelerating though an electric field of 1 volt in a vacuum (1 ev = 1.6×10^{-19} joule).

Element – An atom that has a specific number of protons in its nucleus, which is referred to as the atomic number. Numerous isotopes may occur that have varying numbers of protons. These are distinguished by the atomic weight of the atom. There are ninety-four elements that occur naturally (or as fission products).

External dose of radiation – A radiation dose received from a source external to the body.

Fallout – Radioactive material, mostly fission products, often mixed with soil and debris released into the environment by nuclear explosions. Hundreds of isotopes may be involved.

Fireball – A mass of air surrounding a nuclear explosion and heated to temperatures that may reach tens of millions of degrees Celsius within millionths of a second mostly by absorption of X-rays. Initially, the fireball expands by radiation propagation, but eventually expands as a shock wave. Fireballs rapidly cool but remain visible with temperatures similar to the surface of the Sun near 6000°C for less than a minute.

Firestorm – A stationary, massive fire formed by coalescence of numerous smaller fires with inrushing winds.

First strike – A preemptive surprise attack done because one country believes it is so much more powerful than their adversary that they could eliminate the adversary's potential to retaliate by attacking first.

Fission – A process in which a heavy atom is split by a neutron into lighter elements, which releases energy. Usually ^{235}U and ^{239}Pu are the most important fissionable elements.

Fission fraction – The fraction of the yield of a nuclear weapon that is due to fission.

Fission products – The lighter elements that are produced by fission, which may include 300 isotopes of 36 elements.

Flash burn – A burn caused by exposure of the skin to light radiated by a nuclear explosion (thermal radiation).

Fusion – A process in which lighter elements combine to form heavier elements with a net loss of mass that is used to produce energy.

Gamma ray – Light at frequencies such that the gamma ray photons have energies between a few thousand electron volts and about 10 million electron volts. Gamma rays are produced by nuclear reactions such as fission and radioactive decay.

Giga – A unit prefix equal to 10^9, or 1 billion.

Gigayear (Gyr) – A unit of time that is 10^9 years or, in words, 1 billion years. A Gya, is a billion years ago.

Gray – The absorption of 1 joule of energy per kilogram of matter. The rad is the equivalent in centimeter-gram units. One gray is the same as 100 rads.

Groundburst – A nuclear explosion detonated at ground level, producing a crater and significant fallout but less widespread blast damage than an airburst.

Ground zero – The location on the ground directly below an air burst or above a groundburst.

Gun-type weapon – A type of weapon in which two pieces of fissionable material are brought together very rapidly to form a mass that exceeds the critical mass so that a fission chain reaction occurs. Usually, this type of weapons uses ^{235}U.

Half-life – The time required for the activity of a radioactive material, or the number of atoms of the material, to decrease to half its original value.

Hectopascal – The new metric unit for pressure, which is the same as a millibar (mb), the traditional metric unit for pressure, which is still the conventional unit used by meteorologists (1 hectopascal = 100 pascals).

Height of burst – The height above the ground at which a weapon is detonated. The optimum height of burst is an altitude chosen to maximize some property of the explosion such as the area within which the overpressure exceeds some chosen value.

Hydrogen bomb – A nuclear explosion in which part of the energy comes from fusion.

ICBM – Intercontinental ballistic missile. These are typically land-based missiles that carry one or more nuclear warheads.

Implosion weapon – A type of weapon in which a subcritical mass of fissionable material is compressed by a conventional explosion so that its density becomes great enough to become supercritical. Usually, ^{239}Pu is used in an implosion weapon.

INF – Intermediate Range Nuclear Forces Treaty of 1988, which was signed by Presidents Ronald Reagan and Mikhail Gorbachev. It banned land-based missiles with ranges from 500 to 5500 km. The United States withdrew from the treaty in 2019, which was soon followed by a Russian withdrawal.

Infrared – A range of wavelengths, or colors, of light that lie at larger wavelengths than the red colors and beyond our eyes' ability to see them. All objects with temperatures above absolute zero on the Kelvin scale constantly radiate light, including fires and us. The energy in radiation increases as the fourth power of the temperature of the heat source, so an object that is twice as hot as another radiates sixteen times the amount of light radiation. The wavelength of the emitted light varies in inverse proportion to its temperature. The 6000°C Sun mostly radiates visible light, while objects at terrestrial temperatures mostly radiate infrared light. The wavelength range of infrared light is about 0.7 micrometers (visible red) to 700 micrometers.

Initial nuclear radiation – Radiation from the fireball during the first minute after the explosion. The most dangerous components are neutrons and gamma rays.

Internal dose of radiation – A dose of radiation received from a radioactive material inside the body. The most important sources that accumulate in the body are iodine-131 in the thyroid and strontium-90 and ^{239}Pu in bones. Cesium-137 is especially dangerous following reactor accidents but does not accumulate in the body.

Ionization – The process in which an electron is removed from an atom or molecule leaving behind a positively charged atom or molecule.

Ionizing radiation – High-energy light such as gamma rays or X-rays (or particles such as alpha particles, beta particles, or neutrons moving close to the speed of light) that can ionize atoms or molecules by detaching electrons.

Isotope – Forms of the same element that are usually chemically identical but have different molecular weights because they contain different numbers of neutrons. They may have different half-lives.

Kiloton of energy – A kt is the energy equivalent to the explosion of 1000 tons of the conventional explosive trinitrotoluene, or TNT. Strictly defined as 10^{12} calories, 4.18×10^{19} ergs, or 4.18×10^{12} joules.

Line of Control – The de facto border in Kashmir between India and Pakistan.

Megaton of energy – Megaton, Mt, or the energy equivalent to the explosion of 1 million tons of conventional explosive TNT, which equals 4.18×10^{15} joules.

Megaton of mass – A million metric tons of mass, or a Teragram, Tg.

Micron (micrometer) – A unit of distance that is one-millionth of a meter.

Microsecond – A unit of time that is one-millionth of a second.

Milli – A unit multiplier equal to 0.001 or one-thousandth.

Millisecond – A unit of time that is one-thousandth of a second.

MIRV – Multiple independently-targetable reentry vehicle. A ballistic missile that carries more than one warhead is MIRVed.

Monomer – A single member of a cluster. For example, monomers of black carbon usually are found in the atmosphere in a cluster, as illustrated in Figure 3.6.

Near-Earth object – A comet or asteroid whose orbit allows a closest approach to the Sun being within 1.3 times the distance between the Sun and the Earth.

Neutron – A subatomic particle with the same mass as a proton but with no charge. Hydrogen is the only element that does not have a neutron in its nucleus.

New START – The New Strategic Arms Reduction Treaty, which was signed by Presidents Barack Obama and Dimitri Medvedev in 2010 and extended by Presidents Joseph Biden and Vladmir Putin until February 2026. It does not restrict the numbers of weapons in storage or the number of tactical weapons. For each country, it allows 1550 deployed strategic nuclear warheads, counting each bomber as one warhead. It restricts the sum of deployed and non-deployed launchers (ICBMs, SLBMs, and bombers) to 800 and deployed launchers to 700.

Nuclear radiation – Particles traveling near the speed of light, such as neutrons, alpha particles, and beta particles, and light with small wavelengths so that they carry a lot of energy, such as gamma rays. X-rays are not nuclear radiation since they are not emitted from atomic nuclei. X-rays and gamma rays both have enough energy to be ionizing radiation, so both are dangerous.

Nuclear test – A test of a nuclear weapon. Currently tests above the ground, in outer space, and under the ocean are banned by the Nuclear Test Ban Treaty of 1963, which was signed by the United States, the Soviet Union, and Great Britain. The Comprehensive Nuclear Test Ban Treaty, which would ban all testing of nuclear devices, was passed in the UN General Assembly in 1996, but has not received the signatures needed to be ratified. Nevertheless, most countries have chosen to obey the treaty. The Soviet Union announced a moratorium on testing in 1991, and the US Congress agreed to reciprocate.

Nuclear weapon – An atomic (all fission) bomb or device or a hydrogen (partly fusion) bomb or device. Thermonuclear bombs, also called hydrogen bombs, derive part of their energy from fusion, but much of the energy comes from fission.

Nuclear winter – A substantial reduction in global temperature resulting in a year or more of subfreezing temperatures at midlatitudes in the Northern Hemisphere that is expected to occur from soot generated by burning cities injected into the atmosphere during a nuclear war using a large fraction of the current arsenals. Even a war with only about 1% of the nuclear arsenals, such as between India and Pakistan, could lead to cold enough temperatures to damage agriculture worldwide, leading to mass starvation; however, such cooling is not large enough to be called nuclear winter.

Nucleus – The atomic nucleus is a small central, positively charged region of the atom that contains most of the mass of the atom. It is made of protons and neutrons, which themselves are made of less-massive particles. The number of protons, the atomic number, defines the element, while varying numbers of neutrons define the isotopes of the element and the atomic weight.

Overpressure – Short-duration, excess air pressure encountered in the blast wave of a nuclear explosion (or any other type of shock wave). Overpressure of a few pounds per square inch is sufficient to destroy typical wooden houses.

Planetesimal – Small solid objects like asteroids and comets that are thought to have collided with each other to form Earth and the other planets. The planetesimals themselves are thought to have formed by collisions of dust grains and gas. Collisions of dust grains built pebbles, and collisions of pebbles built planetesimals whose collisions eventually built the planets according to theory.

Proton – One of the constituents of the nucleus of an atom, which defines the atomic number and identifies the element. It carries one unit of positive electrical charge.

Rad – A unit of absorbed dose of radiation. It represents the absorption of 100 ergs (10^{-7} joules) of nuclear (or ionizing) radiation per gram of absorbing material, such as body tissue. A modern equivalent unit is the Gray (1 gray = 100 rads).

Radioactivity – The spontaneous emission of radiation, generally alpha or beta particles or gamma rays.

Radius of destruction – The distance from a nuclear blast within which destruction is near total, often taken as the zone of 5 pounds per square inch overpressure.

Rainout – The removal of particles or gases from the atmosphere by rainfall.

Rational war – A war in which only a small number of weapons are used to demonstrate the determination of the combatants to fight for their goals, but with the expectation that the war will end with minimal damage.

Rem – A unit of biological dose of radiation. The number of rems of radiation is equal to the number of rads absorbed multiplied by the relative biological effectiveness of the given radiation for a specified effect. The rem is also the unit of dose equivalent, which is equal to the product of the number of rads absorbed and the "quality factor" of the radiation. Rem is an older term now replaced by the Sievert.

Roentgen – The amount of ionizing radiation that will produce in 1 kilogram of air an amount of positive or negative charges equal to 2.58×10^{-4} coulomb under standard conditions.

Self-assured destruction – Destruction brought upon the country using a first strike due to the environmental damage caused by fires producing smoke and impacting the climate. It is a form of national suicide.

Shock wave – A pulse or wave propagating through the air, ground, or water initiated by an earthquake, impact, or explosion.

Sievert – Current standard unit of biological dose of radiation (1 Sv = 100 rem).

SLBM – Submarine-launched ballistic missile. Ballistic missiles launched by submarines that often carry one or more nuclear warheads.

Spherule – A spherical particle. In the atmosphere, liquid particles are usually spherical. However, it is unusual for solid particles to be spherical. The shooting stars from the K-Pg Distal layer, also called microkrystites by geologists, were once molten and are now spherical solids. Black carbon particles are often found in sheets or clusters of smaller spherical particles called monomers, which formed in a flame. Figure 3.6 illustrates spherules from the impact the killed the dinosaurs and also shows clusters of spherical black carbon monomers.

START – Strategic Arms Reduction Treaty. START I was agreed to by the United States, the Soviet Union, Belarus, Kazakhstan, and Ukraine under President George Herbert Walker Bush and Mikhail Gorbachev in 1991 and restricted the numbers of warheads and delivery vehicles. It expired in 2009.

Stratosphere – One of the layers of the atmosphere. It begins just above the lowest layer, the troposphere, at an altitude of about 10 km at midlatitudes and 16 km in the tropics. The stratosphere gives way to the mesosphere near 50 km. The stratospheric temperature increases with height, so it is very stable against vertical motions. The stratosphere contains the ozone layer, which protects the surface from hazardous ultraviolet radiation. It does not rain in most of the stratosphere. Some clouds form in the polar night, which cause the formation of the ozone hole by moving chlorine and bromine containing chemicals into ozone reactive compounds, and which precipitate some water and nitrogen compounds out of the stratosphere.

Subduction – A geologic process that occurs when oceanic crust, largely made of dense basaltic rock, is forced under low-density continental crust by continental drift. This process forms deep ocean trenches in many places, and some types of volcanoes.

Surface burst or ground burst – A nuclear explosion whose distance above the ground or water surface is lower than the radius of the fireball. Surface bursts are used to attack buried targets, and they create fallout by lifting soil from the ground. Radiation in the fireball attaches to the soil, which then falls out in a few days or less.

Terminal High Altitude Area Defense (THAAD) – A US missile defense system that uses missiles to collide with incoming short-, medium-, and intermediate-range missiles and destroy them by impact.

Thermal flash – An intense burst of heat radiation in the seconds following a nuclear explosion. The thermal flash of a large weapon can ignite fires and cause third-degree burns many miles from the explosion.

Thermal radiation – Thermal radiation is light that is emitted based on the temperature of the emitting object. Infrared light is generated by objects between room temperature and a few thousand degrees Celsius. Temperatures near 6000°C, the surface temperature of the Sun, generate visible light. Higher temperatures generate ultraviolet light and X-rays.

Thermonuclear – A type of weapon in which very high temperatures are generated by fission reactions, which then cause fusion of light elements to occur.

TNT equivalent – A measure of the energy release in a nuclear explosion expressed as the energy from the explosion of an equivalent mass of trinitrotoluene (TNT). A Mt of TNT is defined to release an energy of 4.18×10^{9} Joules.

Tritium – An isotope of hydrogen with one proton and two neutrons and one electron.

Tropopause – The boundary between the stratosphere and the troposphere, located near 10 km at midlatitudes and 16 km in the tropics.

Troposphere – The lowest layer of the atmosphere, extending from the ground up to the tropopause. The temperature declines with altitude and latitude in the troposphere, which generates efficient mixing throughout the troposphere. Rain washes particles and smoke out quickly here, in a week or two.

Ultraviolet – Light with wavelengths between 0.01 microns where X-rays begin and about 0.4 microns where visible light begins. Unlike visible light, ultraviolet light has enough energy to destroy some molecules, leading to cancers and cataracts in people and to most photochemistry in Earth's atmosphere.

X-ray – Ionizing light radiation with energies between about 100 electron volts and 100,000 electron volts. X-rays are produced when electrons hit a target or when electrons in an atom rearrange themselves. Gamma rays are produced in the atomic nucleus during fission or fusion. Although there is some overlap in energies between gamma rays and X-rays, X-rays usually have much lower energies than gamma rays, so they are not able to penetrate as deeply through tissues or materials such as metal.

Yield – The total energy released in a nuclear explosion. It is usually expressed in terms of the energy from the explosion of an equivalent mass of TNT. Usually units of kilotons (kt) or megatons (Mt) are used.

Acknowledgments

We thank the Open Philanthropy Project for providing funds for our scientific studies and those of our colleagues. The Future of Life Institute has further invested in our work and expanded funding to scientists in other nations. We especially thank Charles Bardeen, Richard Turco, Sherri West, and Frank Von Hippel for providing detailed comments on the entire manuscript and Neil Gordon for helping proofread the final manuscript. We thank Lili Xia, Alan Harris, Jan Smit, Wendy Wolbach, David Kring, Martin Schmeider, and David Morrison for providing figures and photos used in the book, and also thank Rod Nave of the HyperPhysics Project at Georgia State University, Mike Run, the Union of Concerned Scientists, Hans Kristensen of the Federation of American Scientists, and the University Corporation for Atmospheric Research for permission to use their figures in the book. We have had invaluable input from many people, including Lili Xia, Nicole Lovenduski, Cheryl Harrison, Joshua Coupe, Jonas Jägermeyr, Kim Scherrer, Tyler Rohr, and Michael Mills. We especially thank Jeremy Lewis, Kanimozhi Ramamurthy, and Keith Cline at Oxford University Press for helping us with the publication process and editing the manuscript.

References

A'Hearn, M. F., et al., 2005: Deep impact: Excavating Comet Tempel 1, *Science*, **310**, 258–264.

Ahrens, T. J., and A. W. Harris, 1992: Deflection and fragmentation of near-Earth asteroids, *Nature*, **360**, 429–433.

Albright, D., F. Berkhout, and W. Walker, 1997: *Plutonium and Highly Enriched Uranium 1996: World Inventories, Capabilities, and Policies*, New York: Oxford University. Press, 502 pp. (updated http://www.isis-online.org).

Aleksandrov, V. V., and G. L. Stenchikov, 1983: On the modeling of the climatic consequences of the nuclear war, *Proc. Applied Math*, Computing Centre, USSR Academy of Sciences, Moscow, 21 pp.

Alvarez, L., W. Alvarez, F. Asaro, and H. V. Michel, 1980: Extraterrestrial cause for the Cretaceous-Tertiary extinction, *Science*, **208**, 1095–1108.

Alvarez, R., J. Beyea, K. Janberg, J. Kang, E. Lyman, A. Macfarlane, G. Thompson, and F. N. von Hippel, 2003: Reducing the hazards from stored spent power-reactor fuel in the United States, *Science and Global Security*, **11**(1), 1–51.

Alvarez, W., 1997: *T. Rex and the Crater of Doom*, Princeton: Princeton University Press, 208 pp.

AMBIO, 1982: *Nuclear War: The Aftermath*, Oxford: Pergamon Press, 198 pp.

Andreae, M. O., 2019: Emissions of trace gases and aerosols from biomass burning-an updated assessment, *Atmos. Chem. Phys.*, **19**, 8523–8546.

Arakawa, M., et al., 2020: An artificial impact on the asteroid (162173) Ryugu formed a crater in the gravity-dominated regime, *Science*, **368**, 67–71.

Artemieva, N., and J. Morgan, 2020: Global K-Pg layer deposited from a dust cloud, *Geophys. Res. Lett.*, **47**, e2019GL086562.

Baba, M., F. Ogawa, S. Hiura, and N. Asada, 2011: Height estimation of Hiroshima A-bomb mushroom cloud from photos, Chapter 1.6 in *Revisit the Hiroshima A-bomb with Database: Latest Scientific View on Local Fallout and Black Rain*, edited by M. Aoyama and Y. Oochi (Hiroshima City), 55–67.

Bardeen, C. G., R. R. Garcia, O. B. Toon, and A. J. Conley, 2017: On transient climate change at the Cretaceous–Paleogene boundary due to atmospheric soot injections. *Proc. Natl Acad. Sci.* **114**, E7415–E7424.

Bardeen, C. G., D. E. Kinnison, O. B. Toon, M. J. Mills, F. Vitt, L. Xia, J. Jägermeyr, N. S. Lovenduski, K. J. N. Scherrer, M. Clyne, and A. Robock, 2021: Extreme ozone loss following nuclear war resulting in enhanced surface ultraviolet radiation, *J. Geophys. Res. Atmos.*, **126**, e2021JD035079.

Belcher, C. M., R. M. Hadden, G. Rein, J. V. Morgan, N. Artemieva, and T. Goldin, 2015: An experimental assessment of the ignition of forest fuels by the thermal pulse generated by the Cretaceous-Paleogene impact at Chicxulub, *J. Geological Soc.*, **172**, 175–185.

Bohor, B. F., and B. P. Glass, 1995: Origin and diagenesis of K/T impact spherules – from Haiti to Wyoming and beyond, *Meteoritics*, **30**, 182–198.

Boslough, M. B. E., and D. A. Crawford, 1997: Shoemaker-Levy 9 and plume-forming collisions on Earth. *Annals New York Acad. Sci.*, **822**, 236–282.

Bush, B. W., M. A. Dore, G. H. Anno, and R. D. Small, 1991: *Nuclear Winter Source-Term Studies, Volume VI – Smoke produced by a nuclear attack on the United States*, DNA–TR–86–220–V6, Alexandria, VA: Defense Nuclear Agency.

Byron, Lord, 1816: Darkness, in *The Prisoner of Chillon and Other Poems*, London: John Murray, 27-31.

Center for Near Earth Object Studies, https://cneos.jpl.nasa.gov.

Cheng, A. F. et al., 2023: Momentum transfer from the DART mission kinetic impact on asteroid Dimorphos, *Nature*, **616**, 457–460.

Chervin, R., 2024: *The Cold War in the Himalayas; Multinational Perspectives on the Sino-Indian Border Conflict, 1950-1970*, Amsterdam: Amsterdam University Press, 294 pp.

Chyba, C. F., P. J. Thomas, and K. J. Zahnle, 1993: The 1908 Tunguska explosion: atmospheric disruption of a stony asteroid, *Nature*, **361**, 40–44.

Clodfelter, M., 1992: *Warfare and Armed Conflicts: A Statistical Reference to Casualty and other Figures, 1618-1991, Vol II 1900-1991*, McFarland and Co, 1414 pp.

Coupe, J., C. G. Bardeen, A. Robock, and O. B. Toon, 2019: Nuclear winter responses to global nuclear war in the Whole Atmosphere Community Climate Model Version 4 and the Goddard Institute for Space Studies ModelE. *J. Geophys. Res. Atmos.*, **124**, 8522–8543.

Coupe, J., S. Stevenson, N. S. Lovenduski, T. Rohr, C. S. Harrison, A. Robock, H. Olivarez, C. G. Bardeen, and O. B. Toon, 2021: Nuclear Niño response observed in simulations of nuclear war scenarios. *Communications Earth & Environment*, **2**, 18.

Covey, C., S. H. Schneider, and S. L. Thompson, 1984: Global atmospheric effects of massive smoke injections from a nuclear war: Results from general circulation model simulations, *Nature*, **308**(5954), 21–25.

Crutzen, P. J., and J. W. Birks, 1982: The atmosphere after a nuclear war: Twilight at noon, *Ambio*, **11**, 114–125.

DePalma, R. A., J. Smit, D. A. Burnham, K. Kuiper, P. L. Manning, A. Oleinik, P. Larson, F. J. Maurrasse, J. Vellekoop, M. A. Richards, L. Gurche, and W. Alvarez, 2019: A seismically induced onshore surge deposit at the K-Pg boundary, North Dakota, *Proc. Nat. Acad. Sci.*, **116**, 8190–8199.

Department of Defense, 2018: *Base Structure Report-Fiscal year 2018; Baseline, A Summary of the real property inventory data*, https://www.acq.osd.mil/eie/Downloads/BSI/Base%20Structure%20Report%20FY18.pdf

Dutton, E. G., and J. R. Christy, 1992: Solar radiative forcing at selected locations and evidence for global lower tropospheric cooling following the eruptions of El Chichón and Pinatubo, *Geophys. Res. Lett.*, **19**, 2313–2316.

Earth Policy Institute Data Center, World grain consumption and stocks as days of consumption, 1960-2012, available at www.earth-policy.org/data_center/C24.

Eden, L., 2004: *Whole World on Fire: Organizations, Knowledge, & Nuclear Weapons Devastation*, Ithaca, N.Y.: Cornell University Press, 365 pp.

Ehrlich P., et al., 1983: Long term biological consequences of nuclear war, *Science*, **222**, 1293–1300.

Ellsberg, D., 2017: *The Doomsday Machine: Confessions of a Nuclear War Planner*, New York: Bloomsbury, 420 pp.

Fears, T. R., J. Scotto, and M. A. Schneiderman, 1976: Skin cancer, melanoma and sunlight, *Am. J. Public Health*, **66**, 461–464.

Fromm, M., R. Servranckx, B. J. Stocks, and D. A. Peterson, 2022: Understanding the critical elements of the pyrocumulonimbus storm sparked by high-intensity wildland fire, *Comm. Earth Environ.*, **3**, 243.

Glasstone, S., and P. J. Dolan, 1977: *The Effects of Nuclear Weapons, 3rd ed.*, Washington, D.C.: United States Department of Defense and Energy Research and Development Administration, 653 pp.

Goderis, S., et al., 2021: Globally distributed iridium layer preserved within the Chicxulub impact structure, *Sci. Adv.*, **7**, eabe3647.

Goldin, T. J., and H. J. Melosh, 2009: Self-shielding of thermal radiation by Chicxulub impact ejecta: Firestorm or fizzle?, *Geology*, **37**, 1135–1138.

Green, T., P. R. Renne, and C. B. Keller, 2022: Continental flood basalts drive the Phanerozoic extinctions, *Proc. Nat. Acad. Sci.*, **119**(38), e2120441119.

Gulick, S. P. S., 2019: The first day of the Cenozoic, *Proc. Nat. Acad. Sci.*, **116**, 19,342–19,351.

Haberle, R. M., C. Leovy, and J. B. Pollack, 1982: Some effects of global dust storms on the atmospheric circulation of Mars, *Icarus*, **50**, 322–367.

Hale, Sir Lonsdale Augustus, 1896: *The Fog of War*, London: Edward Stanford, 17 pp.

Hansen, G., and E. Condon, 1989: *Denial of Disaster; The Untold Story and Photographs of the San Francisco Earthquake and Fire of 1906*, Cameron & Company, 160 pp.

Harris, A. W., and P. W. Chodas, 2021: The population of near-earth asteroids revisited and updated, *Icarus*, **365**, 114452.

Harris, N. W., and D. W. Hughes, 1994: Asteroid-Earth collision velocities, *Planet. Space Sci.*, **42**, 285–289.

Harrison, C. S., et al., 2022: A new ocean state after nuclear war, *AGU Advances*, **3**, e2021AV000610.

Harwell M. A., and T. C. Hutchinson, 1985: *Environmental Consequences of Nuclear War, Vol. 2, Ecological and Agricultural Effects*, SCOPE-28, Chichester, England: Wiley, 526 pp.

Helfand, I., L. Forrow, M. McCally, and R. K. Musil, 2002: Projected US casualties and destruction of US medical services from attacks by Russian nuclear forces, *Medicine and Global Survival*, **7**, 68–76.

Hertsgaard, M., 2000: Mikhail Gorbachev explains what's rotten in Russia, *Salon.com*, https://www.salon.com/2000/09/07/gorbachev/

Houghton, R. A., 2005: Above ground forest biomass and the global carbon balance, *Global Change Biology*, **11**, 945–958.

IAEA (International Atomic Energy Agency), 2008: *Estimation of Global Inventories of Radioactive Waste and Other Radioactive Materials*, IAEA-TECDOC-1591, Vienna: IAEA, 45 pp.

International Panel on Fissile Materials, Fissile material stocks, http://fissilematerials.org/ Downloaded April 19, 2024.

Ishikawa, E., and D. L. Swain (Translators), 1981: *Hiroshima and Nagasaki: The Physical, Medical, and Social Effects of the Atomic Bombings*, New York: Basic Books, 702 pp.

Jablonski, D., 1994: Extinctions in the fossil record, *Phil. Trans. R. Soc. Lond. B*, **344**, 11–17.

Jacobsen, A., 2024: *Nuclear War – A Scenario*, New York: Dutton, 373 pp.

Jägermeyr, J., et al., 2020: A regional nuclear conflict would compromise global food security. *Proc. Nat. Acad. Sci.*, **117**, 7071–7081.

Jet Propulsion Laboratory, 2024: Fireballs, https://cneos.jpl.nasa.gov/fireballs/

Johnson, B. C., and H. J. Melosh, 2012: Formation of spherules in impact produced vapor plumes, *Icarus*, **217**, 416–430.

Kemp, R. S., 2014: The nonproliferation emperor has no clothes: The gas centrifuge, supply-side controls and the future of nuclear proliferation, *International Security*, **38**, 39–78.

Khan, H. F., and R. W. French, 2013: *South Asian Stability Workshop: A crisis simulation exercise* (Naval Postgraduate School Center on Contemporary Conflict Project on Advanced Systems and Concepts for Countering WMD Report Number 2013–008).

Kring, D., and M. Boslough, 2014: Chelyabinsk: Portrait of an asteroid airburst, *Physics Today*, **67**(9), 32–37.

Kristensen, H. M., and M. Korda, 2019: Tactical nuclear weapons, 2019, Bull, Atomic Scientists, 75(5), 252–261, doi 10.1080/00963402.2019.1654273.

Kristensen, H. M., and M. Korda, 2021: United Kingdom nuclear weapons, *Bulletin of the Atomic Scientists*, **77**(3), 153–158.

Kristensen, H. M., and M. Korda, 2022a: Indian nuclear weapons, *Bulletin of the Atomic Scientists*, **78**(4), 224–236.

Kristensen, H. M., and M. Korda, 2022b: North Korean nuclear weapons, *Bulletin of the Atomic Scientists*, **78**(5), 273–294.

Kristensen, H. M., M. Korda, and E. Johns, 2023a: French nuclear weapons, *Bulletin of the Atomic Scientists*, **79**(4), 272–281.

Kristensen, H. M., M. Korda, and E. Johns, 2023c: Pakistan nuclear weapons, *Bulletin of the Atomic Scientists*, **79**(5), 329–345.

Kristensen, H. M., M. Korda, E. Johns, and M. Knight, 2024a: United States nuclear weapons, 2024: *Bulletin of the Atomic Scientists*, **80**(3), 182–208.

Kristensen, H. M., M. Korda, E. Johns, and M. Knight, 2024b: Russian nuclear weapons, 2024: *Bulletin of the Atomic Scientists*, **80**(2), 118–145.

Kristensen, H. M., M. Korda, E. Johns, and M. Knight, 2024c: Estimated global nuclear warhead inventories, Federation of American Scientists, https://fas.org/initiative/status-world-nuclear-forces/.

Kristensen, H.M., M. Korda, and E. Reynolds, 2023b: Chinese nuclear weapons, *Bulletin of the Atomic Scientists*, **79**(2), 108–133.

Kunkle, T., and B. Ristet, 2013: *Castle Bravo: Fifty years of legend and lore, A guide to off-site radiation exposures*, Defense Threat Reduction Information Analysis Center, DTRIAC SR-12-001, LA-UR-04-1400.

Larson, D. A., and R. D. Small, 1982: *Analysis of the Large Scale Urban Fire Environment, Part II. Parametric Analysis and Model City Simulations*, Pacific-Sierra Res. Corp., Contract EMW-C-0747, Work Unit 2564E.

Lawton, J. H., and R. M. May, 1995: Assessing extinction rates, in J. H. Lawton and R. M. May (eds.) *Extinction Rates*, Oxford University Press, 1–24.

Lewis, K. N., 1979: The prompt and delayed effects of nuclear war, *Scientific American*, **241**, 35–47.

Lieber, K. A., and D. G. Press, 2006: The rise of U.S. nuclear primacy, *Foreign Affairs*, https://www.foreignaffairs.com/articles/united-states/2006-03-01/rise-us-nuclear-primacy.

London, J., 1906: Story of an eyewitness, *Collier's, The National Weekly*, New York: Colliers, May 5, 1906.

Longrich, N, R., J. Scriberas, and M. A. Willis, 2016: Severe extinction and rapid recovery of mammals across the Cretaceous-Paleogene boundary, and the effects of rarity on patterns of extinction and recovery, *J. Evolutionary Biology*, **29**, 1495–1512.

Lovenduski, N. S., C. S. Harrison, H. Olivarez, C. G. Bardeen, O. B. Toon, J. Coupe, A. Robock, T. Rohr, and S. Stevenson, 2020: The potential impact of nuclear conflict on ocean acidification, *Geophys. Res. Lett.*, **47**, e2019GL086246.

Lubin, P., 2023: Terminal planetary defense, https://arxiv.org/abs/2110.07559.

Lubin, P., and A. N. Cohen, 2022: Don't forget to look up, https://arxiv.org/abs/2201.10663v4.

MacCracken, M. C., 1983: Nuclear war: Preliminary estimates of the climatic effects of a nuclear exchange, Lawrence Livermore National Laboratory Report UCRL-89770 (Livermore, CA).

Mark, J. C., 1993: Explosive properties of reactor-grade plutonium, *Science & Global Security*, **4**, 111–128.

McKinzie, M., Z. Mian, A. H. Nayyar, and M. V. Ramana, 2001a: The risks and consequences of nuclear war in South Asia, in S. Kothari and Z. Mian (eds.) *Out of the Nuclear Shadow*, London: Zed Books Ltd., 185–196.

McKinzie, M. G., T. B. Cochran, R. S. Norris, and W. A. Arkin, 2001b: *The U.S. Nuclear War Plan: A Time for a Change*, Washington, D.C.: National Resources Defense Council, 208 pp.

Melosh, H. J., and I. V. Nemchinov, 1993: Solar asteroid diversion, *Nature*, **366**, 21–22.

Melosh, H. J., N. M. Schneider, K. J. Zahnle, and D. Latham, 1990: Ignition of global wildfires at the Cretaceous–Tertiary boundary, *Nature*, **343**, 251–254.

Middleton, H., 1982: Epidemiology: The future is sickness and death. *Ambio* 11, 100–105.

Mills, M. J., O. B. Toon, J. Lee-Taylor, and A. Robock, 2014: Multi-decadal global cooling and unprecedented ozone loss following a regional nuclear conflict. *Earth's Future*, **2**, 161–176.

Mills, M. J., O. B. Toon, R. P. Turco, D. E. Kinnison, and R. R. Garcia, 2008: Massive global ozone loss predicted following regional nuclear conflict. *Proc. Natl. Acad. Sci.*, **105**, 5307–5312.

Milne, D. H., and C. McKay, 1982: Response of marine plankton communities to a global atmospheric darkening, in L. T. Silver and P. H. Schultz (eds.) *Geological Implications of Impacts of Large Asteroids and Comets on the Earth, Geol. Soc. Am. Spec. Pap.*, **190**, 297–303.

Morgan, J., C. Lana, A. Kearsley, B. Coles, C. Belcher, S. Montanari, E. Díaz-Martínez, A. Barbosa, and V. Neumann, 2006: Analysis of shocked quartz at the global K-P boundary indicate an origin from a single, high-angle, oblique impact at Chicxulub, *Earth Planetary Sci. Lett.*, **251**, 264–279.

Morgan, J. V., T. J. Bralower, J. Brugger, and K Wünnemann, 2022: Chicxulub impact crater formation and environmental consequences, *Nature Reviews*, **3**, 338–354.

National Research Council, 1985: *The Effects on the Atmosphere of a Major Nuclear Exchange.* Washington, D.C.: The National Academies Press, 206 pp.

National Safety Council, 2024: Odds of dying, 2021, https://injuryfacts.nsc.org/all-injuries/preventable-death-overview/odds-of-dying/.

NEA (Nuclear Energy Agency), 2002: The release, dispersion, deposition and behaviour of radionuclides, Chapter II of *Chernobyl: Assessment of Radiological and Health Impacts*, Paris: OECD Publishing, 33–51.

Oreskes, N., and E. M. Conway, 2010: *Merchants of Doubt: How a Handful of Scientists Obscured the Truth on Issues from Tobacco Smoke to Global Warming*, New York: Bloomsbury Press, 368 pp.

OTA, 1979: *The Effects of Nuclear War*, Washington, D.C.: Office of Technology Assessment, NTIS order #PB-296946, 154 pp.

Oughterson, A. W., and S. Warren, 1956: *Medical Effects of the Atomic Bomb in Japan*, New York: McGraw-Hill, 477 pp.

Pausata, F. S. R., J. Lindvall, A. M. L. Ekman, and G. Svensson, 2016: Climate effects of a hypothetical regional nuclear war: Sensitivity to emission duration and particle composition. *Earth's Future*, **4**(11), 498–511.

Perry, W. J., and T. Z. Collina, 2019: *The Button: The new nuclear arms race and presidential power from Truman to Trump*, Dallas: BenBella Books, 280 pp.

Philippe, S., 2023: Who would take the brunt of an attack on U.S. nuclear missile silos?, *Scientific American*, **329**(5), 46.

Pittock, A. B., T. P. Ackerman, P. J. Crutzen, M. C. MacCracken, C. S. Shapiro, and R. P. Turco, 1989: *Environmental Consequences of Nuclear War, SCOPE 28, Vol 1. Physical and Atmospheric Effects*, 2nd ed., New York: John Wiley & Sons, 428 pp.

Pompeo, M., 2023: *Never Give an Inch: Fighting for the America I Love*, New York: Broadside Books, 464 pp.

Popova, O. P., et al., 2013: Chelyabinsk airburst, damage assessment, meteorite recovery and characterization, *Science*, **342**, 1069–1073.

Richard, T. T., 2016: Nuclear weapons targeting: The evolution of law and U.S. policy, *Military Law Review*, **224**, 862–978.

Ritchie, H., and M. Poser, 2018: Causes of Death, https://ourworldindata.org/causes-of-death.

Rivkin, A. S., et al., 2021: The Double Asteroid Redirection Test (DART): Planetary defense investigations and requirements, *Planet. Sci. J.*, **2**, 173.

Robertson, D. S., W. M. Lewis, P. M. Sheehan, and O. B. Toon, 2013: K-Pg extinction patterns in marine and freshwater environments: The impact winter model, *J. Geophys. Res. Biogeosci.*, **118**, 1006–1014.

Robertson, D. S., M. C. McKenna, O. B. Toon, S. Hope, and J. A. Lillegraven, 2004: Survival in the first hours of the Cenozoic, *GSA Bulletin* (May/June 2004), **116**, 760–768.

Robock, A., 1984: Snow and ice feedbacks prolong effects of nuclear winter, *Nature*, **310**, 667–670.

Robock, A., 1988: Enhancement of surface cooling due to forest fire smoke. *Science*, **242**, 911–913.

Robock, A., 1991: Surface cooling due to forest fire smoke, *J. Geophys. Res.*, **96**, 20,869–20,878.

Robock, A., 2020: Benefits and risks of stratospheric solar radiation management for climate intervention (geoengineering). *The Bridge*, **50**, 59–67.

Robock, A., L. Oman, G. L. Stenchikov, O. B. Toon, C. Bardeen, and R. P. Turco, 2007a: Climatic consequences of regional nuclear conflicts. *Atm. Chem. Phys.*, **7**, 2003–2012.

Robock, A., L. Oman, and G. L. Stenchikov, 2007b: Nuclear winter revisited with a modern climate model and current nuclear arsenals: Still catastrophic consequences. *J. Geophys. Res.*, **112**, D13107.

Robock, A., and O. B. Toon, 2012: Self-assured destruction: The climate impacts of nuclear war, *Bull. Atomic Scientists*, **68**(5), 66–74.

Robock, A., and B. Zambri, 2018: Did smoke from city fires in World War II cause global cooling? *J. Geophys. Res. Atmos.*, **123**, 10,314–10,325.

Russell, D., 1979: The enigma of the extinction of the dinosaurs, *Ann. Rev. Earth Planet. Sci.*, **7**, 163–182.

Russell, D. A., and R. Séguin, 1982: Reconstruction of the small Cretaceous theoropod Stenonychosaurus inequalis and a hypothetical dinosauroid, *Syllogeous*, **37**, 1–43.

Sagan, C., 1980: *Cosmos*, Random House, 365 pp.

Sagan, C., and S. J. Ostro, 1994: Dangers of asteroid deflection, *Nature*, **368**, 501.

Sagan, C., and R. Turco, 1990: *A Path Where No Man Thought – Nuclear Winter and the End of the Arms Race*, New York: Random House, 499 pp.

Scherrer, K. J. N., et al., 2020: Marine wild-capture fisheries after nuclear war. *Proc. Nat. Acad. Sci.*, **117** (47), 29,748–29,758.

Schmieder. M., and D. A. Kring, 2020: Earth's impact events through geologic time: A list of recommended ages for terrestrial impact structures and deposits, *Astrobiology*, **20**, 91–141.

Schneider, S. H., and L. E. Mesirow, 1976: *The Genesis Strategy: Climate and Global Survival*, New York: Plenum Press, 419 pp.

Schulte, P., et al., 2010: The Chicxulub asteroid impact and mass extinction at the Cretaceous-Paleogene boundary, *Science*, **327**, 1214–1218.

Segura T. L., K. Zahnle, O. B. Toon, and C. P. McKay, 2013: The effects of impacts on the climates of terrestrial planets, *Comparative Climatology of Terrestrial Planets* (S. J. Mackwell et al., eds.), 417–437, Tucson: University of Arizona Press.

Sepkoski Jr., J. J., 1996: Patterns of Phanerozoic extinction: a perspective from global data bases, in O. H. Waller (ed.), *Global Events and Event Stratigraphy in the Phanerozoic*, Berlin: Springer-Verlag.

Shklovskii, I. S., and C. Sagan, 1966: *Intelligent Life in the Universe*, New York: Dell, 509 pp.

Shoemaker, E. M., R. F. Wolfe, and C. S. Shoemaker, 1990: Asteroid and comet flux in the neighborhood of Earth, in *Global Catastrophes in Earth History*, edited by V. Sharpton and P. Ward, *Spec. Pap. Geol. Soc. Am.*, **247**, 155–170.

Shultz, G. P., W. J. Perry, H. A. Kissinger, and S. Nunn, 2007: A World Free of Nuclear Weapons. *Wall Street Journal*, January 4.

Shultz, G. P., W. J. Perry, H. A. Kissinger, and S. Nunn, 2008: Toward a Nuclear-Free World, *Wall Street Journal*, January 15.

Shultz, G. P., W. J. Perry, H. A. Kissinger, and S. Nunn, 2010: How to Protect Our Nuclear Deterrent, *Wall Street Journal*, January 19.

Shultz, G. P., W. J. Perry, H. A. Kissinger, and S. Nunn, 2011: Deterrence in the Age of Nuclear Proliferation, *Wall Street Journal*, March 7.

Sikkink, P. G., D. C. Lutes, and R. E. Keane, 2009: *Field guide for identifying fuel loading models*, Gen. Tech. Rep. RMRS-GTR-225, Fort Collins, CO: U.S. Department of Agriculture, Forest Service, Rocky Mountain Research Station, 33 pp.

Simonett, D. S., T. N. Barrett, S. Gopal, F. J. Holsmuller, and H. Veregin, 1998: Estimates of the magnitude and spatial distribution of combustible materials in urban areas: A case study of the San Jose Area, California, *Fire and Materials*, **12**, 95–108.

Sleep N. H., K. J. Zahnle, J. F. Kasting, and H. J. Morowitz, 1989: Annihilation of ecosystems by large asteroid impacts on the early Earth. *Nature*, **342**, 139–142.

Smit, J., 1999: The global stratigraphy of the Cretaceous-Tertiary boundary impact ejecta, *Annual Rev. Earth. Planet. Sci.*, **27**, 75–113.

Starr, S., 2023: *Nuclear High-Altitude Electromagnetic Pulse A Mortal Threat to the U.S. Power Grid and U.S. Nuclear Power Plants*, Rethink Government, 188 pp.

Steel, D., 1998: Distributions and moments of asteroid and comet impact speeds upon the Earth and Mars, *Planet Space Sci.*, **46**, 473–478.

Stenchikov, G. L., I. Kirchner, A. Robock, H.-F. Graf, J. C. Antuña, R. G. Granger, A. Lambert, and L. W. Thomason, 1998: Radiative forcing from the 1991 Mt. Pinatubo volcanic eruption, *J. Geophys. Res.*, **103**, 13,837–13,857.

Stenke, A., C. R. Hoyle, B. Luo, E. Rozanov, J. Gröbner, L. Maag, S. Brönnimann, and T. Peter, 2013: Climate and chemistry effects of a regional scale nuclear conflict, *Atmos. Chem. Phys.*, **13**, 9713–9729.

Sternberg, T., 2013: Chinese Drought, Wheat, and the Egyptian Uprising: How a Localized Hazard Became Globalized, in C. E. Werrell and F. Femia (eds.) *The Arab Spring and Climate Change: A Climate and Security Correlations Series*, Center for American Progress. http://americanprogress.org/issues/security/report/2013/02/28/54579/the-arab-spring-and-climate-change/

Tabor, C. R., C. G. Bardeen, B. L. Otto-Bliesner, R. R. Garcia, and O. B. Toon, 2020: Causes and climatic consequences of the impact winter at the Cretaceous-Paleogene boundary. *Geophys. Res. Lett.*, **47**, e60121.

Tanaka, S., and S. Kado, 2015: Analysis of Radioactive Release from the Fukushima Daiichi Nuclear Power Station. Chapter 3 in J. Ahn, et al. (eds), *Reflections on the Fukushima Daiichi Nuclear Accident*, Heidelberg: Springer Cham, 51–83.

Terzian, Y., and E. M. Bilson, eds., 1997: *Carl Sagan's Universe*, Cambridge, UK: Cambridge University Press, 282 pp.

Toon, O. B., C. G. Bardeen, A. Robock, L. Xia, H. Kristensen, M. McKinzie, R. J. Peterson, C. Harrison, N. S. Lovenduski, and R. P. Turco, 2019: Rapid expansion of nuclear arsenals by Pakistan and India portends regional and global catastrophe. *Science Advances*, **5**, eaay5478.

Toon, O. B., J. B. Pollack, T. P. Ackerman, R. P. Turco, C. P. McKay, and M. S. Liu, 1982: Evolution of an impact generated dust cloud and its effects on the atmosphere, *Geological Implications of Impacts of Large Asteroids and Comets on the Earth*, edited by L. T. Silver and P. H. Schultz, *Spec. Pap. Geol. Soc. Am.*, **190**, 187–200.

Toon, O. B., A. Robock, M. Mills, and L. Xia, 2017: Asia treads the nuclear path, unaware that self-assured destruction would result from nuclear war. *J. Asian Studies*, **76**, 437–456.

Toon, O. B., A. Robock, and R. P. Turco, 2008: Environmental consequences of nuclear war, *Physics Today*, **61**(12), 37–42.

Toon, O. B., R. P. Turco, A. Robock, C. Bardeen, L. Oman, and G. L. Stenchikov, 2007: Atmospheric effects and societal consequences of regional scale nuclear conflicts and acts of individual nuclear terrorism, *Atmos. Chem. Phys.*, **7**, 1973–2002.

Toon, O. B., K. Zahnle, D. Morrison, R. P. Turco, and C. Covey, 1997: Environmental perturbations caused by the impacts of asteroids and comets, *Rev. Geophys.*, **35**, 41–78.

Turco, R. P., O. B. Toon, T. P. Ackerman, J. B. Pollack, and C. Sagan, 1983: Nuclear winter: Global consequences of multiple nuclear explosions, *Science*, **222**, 1283–1292.

Turco, R. P., O. B. Toon, T. P. Ackerman, J. B. Pollack, and C. Sagan, 1990: Climate and smoke: An appraisal of nuclear winter, *Science*, **247**, 166–176.

Tyrrell, T., A. Merico, and D. I. A. McKay, 2015: Severity of ocean acidification following the end-Cretaceous asteroid impact, *Proc. Nat. Acad. Sci.*, **112**, 6556–6561.

United Nations, 2017: *World Mortality 2017, Data Booklet* (ST/ ESA/SER.A/412), New York: United Nations Department of Economic and Social Affairs, Population Division, 24 pp.

UNSCEAR (United Nations Scientific Committee on the Effects of Atomic Radiation), 1993: *Sources and Effects of Ionizing Radiation: United Nations Scientific Committee on the Effects of Atomic Radiation 1993 Report to the General Assembly*, United Nations Publication Sales No.E.94.IX.2, 922 pp.

UNSCEAR (United Nations Scientific Committee on the Effects of Atomic Radiation), 2000: *Sources and Effects of Ionizing Radiation: United Nations Scientific Committee on the Effects of Atomic Radiation 2000 Report, Volume I, Annex C*, United Nations Publication Sales No. E.00.IX.3, 291 pp.

Urey, H. C., 1973: Cometary collisions and geologic periods, *Nature*, **242**, 32–33.

Vonnegut Jr., K., 1969: *Slaughterhouse-Five or the Children's Crusade*, New York: Delacorte, 190 pp.

Ward, P. D., and D. Brownlee, 2000: *Rare Earth: Why complex life is uncommon in the universe*, New York: Springer-Verlag, 372 pp.

Wellerstein, A., 2020: Counting the dead at Hiroshima and Nagasaki. *Bull. Atomic Scientists*, https://thebulletin.org/2020/08/counting-the-dead-at-hiroshima-and-nagasaki/

White, C. S., I. G. Bowen, D. R Richmond, and R. L. Corsbie, 1960: *Comparative nuclear effects of biomedical interest*, U. S. Atomic Energy Commission, Civil Effects Study, CEX–58.8.

Williams, D., A. Kuipers, C. Mukai, and R. Thirsk, 2009: Acclimation during space flight: Effect on human physiology, *Canadian Medical Association J.*, **180**, 1317–1323.

Witmer, S., 2017: Nuclear close calls, https://www.wagingpeace.org/nuclear-close-calls/

Witze, A., 2020: How a small nuclear war would transform the entire planet. *Nature*, **579**, 485–487.

Wolbach, W. S., I. Gilmour, E. Anders, C. J. Orth, and R. R. Brooks, 1988: Global fire at the Cretaceous-Tertiary boundary, *Nature*, **334**, 665–669.

Wolbach W. S., I. Gilmour, and E. Anders, 1990: Major wildfires at the Cretaceous/Tertiary boundary, *Geol. Soc. Am. Spec. Pap.*, **247**, 391–400.

World Health Assembly 40, 1987: Effects of nuclear war on health and health services: Report of the WHO Management Group on follow-up of resolution WHA36.28, Appendix 4.b. Geneva: World Health Organization.

Xia, L., A. Robock, K. Scherrer, C. S. Harrison, B. L. Bodirsky, I. Weindl, J. Jägermeyr, C. G. Bardeen, O. B. Toon, and R. Heneghan, 2022: Global food insecurity and famine from reduced crop, marine fishery and livestock production due to climate disruption from nuclear war soot injection. *Nature Food*, **3**, 586–596.

Yu, P., et al., 2019: Black carbon lofts wildfire smoke high into the stratosphere to form a persistent plume. *Science*, **365**, 587–590.

Yu, P. et al., 2021: Persistent stratospheric warming due to 2019-2020 Australian wildfire smoke. *Geophys. Res. Lett.*, **48**, e2021GL092609.

Index